Climate Mitigation in the Land Sector of the Global South

This book challenges dominant, top-down approaches to climate mitigation in the land sector, arguing that without genuine negotiation, recognized land rights, and tailored capacity building for Indigenous Peoples and Local Communities (IPLCs), most solutions will inevitably fail or do harm.

With civil society increasingly acting as a driving force for transparency and accountability around the world, the possibility of achieving feasible, impactful Nature-based Solutions (NbS) is growing. This vision echoes Brazil's call for a mutirão – a collective, bottom-up mobilization being championed at COP30 in Belém. Drawing on decades of fieldwork and critical analysis, Michael Brown reveals why prevailing strategies – from REDD+ to other NbS – remain implausible and risk compounding poverty and marginalization in the Global South. Instead, Brown proposes a paradigm shift, elevating a governance innovation: Negotiated Governance Platforms (NGPs) that bring IPLCs, governments, and climate financiers together as mutually interested partners. Through a rich set of case studies that include successes and failures, he demonstrates how social license, tenure security, and robust capacity building are essential for achieving climate action that will be sustainable because it works for key actors at multiple levels.

Rejecting technocratic quick fixes, this book offers a pragmatic blueprint for moving beyond rhetoric to operational reality, even in contexts characterized by corruption or weak governance. It is essential reading for students, scholars, and practitioners in climate change, natural resource management, conservation, and green economics who seek grounded, actionable strategies that benefit both people and planet.

Michael I. Brown is a natural resources management, community conservation, sustainable agriculture, and resiliency specialist and author, with 40 years of experience designing, implementing, and evaluating projects in multiple sectors in more than 30 countries.

Routledge Advances in Climate Change Research

Resolving the Climate Crisis
US Social Scientists Speak Out
Edited by Kristin Haltinner and Dilshani Sarathchandra

Climate Politics in Populist Times
Climate Change Communication Strategies in Germany, Spain, and Austria
Mirjam Gruber

Confronting Climate Coloniality
Decolonizing Pathways for Climate Justice
Edited by Farhana Sultana

The Cultural Complexity of Carbon
Green Transformations in Contemporary Society
Edited by Steffen Dalsgaard, Andy Lautrup, Katinka A. Schyberg, and Ingmar Lippert

Climate Action and Civil Society in Turkey and Germany
Contested Common Ground
Hande Paker

An Interdisciplinary Approach to Climate Change
Perspectives from the Social Sciences and Humanities
Edited by Andrea Catellani, Louise-Amélie Cougnon, Øyvind Gjerstad, and Armelle Nugier

Climate Mitigation in the Land Sector of the Global South
Making it Work for People and Planet
Michael I. Brown

For more information about this series, please visit: www.routledge.com/Routledge-Advances-in-Climate-Change-Research/book-series/RACCR

Climate Mitigation in the Land Sector of the Global South

Making it Work for People and Planet

Michael I. Brown

Routledge
Taylor & Francis Group
LONDON AND NEW YORK

First published 2026
by Routledge
4 Park Square, Milton Park, Abingdon, Oxon OX14 4RN

and by Routledge
605 Third Avenue, New York, NY 10158

Routledge is an imprint of the Taylor & Francis Group, an informa business

British Library Cataloguing-in-Publication Data
A catalogue record for this book is available from the British Library

ISBN: 978-1-032-46161-8 (hbk)
ISBN: 978-1-032-46163-2 (pbk)
ISBN: 978-1-003-38032-0 (ebk)

DOI: 10.4324/9781003380320

Typeset in Times New Roman
by Apex CoVantage, LLC

Contents

Figures

Tables

Boxes

Acronyms and Abbreviations

AIIB	Asian Infrastructure Investment Bank
AFOLU	Agriculture, Forestry, Other Land Uses
AI	Artificial Intelligence
AR6	Sixth Assessment Report (AR6) of the Intergovernmental Panel on Climatc Change (IPCC)
ASALs	Arid and Semi-Arid Lands
BAU	Business as Usual
BINGOs	Big Conservation (and Development) NGOs
CBD	Convention on Biodiversity
CBDR	Common but Differentiated Responsibilities
CDR	Carbon Dioxide Removal
CI	Conservation International
CIF	Climate Investment Fund
CIFOR	Center for International Forestry
CMLSGS	Climate Mitigation in the Land Sector of the Global South
COP	Conference of Parties
CO2	Carbon dioxide
DRC	Democratic Republic of Congo
EK-JERP	East Kalimantan Jurisdictional Emissions Reduction Program
ER	Emissions Reductions
FMU	Forest Management Unit
FPIC	Free, Prior, and Informed Consent
FPP	Forest Peoples Programme
G77	A coalition of developing countries within the United Nations (UN) that was established in 1964 to promote their collective economic interests and enhance their negotiating power on global issues including climate change
GATC	Global Alliance of Territorial Communities
GBF	Global Biodiversity Framework
GCA	Global Center on Adaptation
GEF	Global Environment Facility
GELTS	Good Enough Land Tenure Security
GtCO2eq y	Gigatonnes of Carbon Dioxide Equivalent Per Year
HPPSS	High Probability Pathway to Achieve Sustainable Solutions

IC	International Community of Donors and Conservation and Development Partners
ICDP	Integrated Conservation and Development Projects
ICI	Inclusive Conservation Initiative
ICRAF	International Center for Agroforestry
IFC	International Finance Corporation
IK	Indigenous Knowledge
IPs	Indigenous Peoples
IPCC	Intergovernmental Panel on Climate Change
IPLCs	Indigenous People and Local Communities
IRM	Innovative Resources Management
JA	Jurisdictional Approach
LDCs	Least Developed Countries
LMICs	Low- and Medium-Income Countries
LTS	Land Tenure Security
LUP	Land-Use Planning
LUPs	Land-Use Plans
MRV	Measurement, Reporting, Verification
NACSO	Namibian Association of CBNRM Support Organizations
NbS	Nature-based Solutions
NBIs	Nature-Based Interventions
NbCS	Nature-based Climate Solutions
NCQG	New Collective Quantified Goal
NCS	Natural Climate Solutions
NDCs	Nationally Determined Contributions
NGPs	Negotiated Governance Platforms
NRT	Northern Rangelands Trust
ODA	Official Development Assistance
PCBRA	Participatory Cost-Benefit and Risk Analysis
PES	Payment for Environmental (or Ecosystem) Services
PPP	Public-Private Partnerships
REDD+	Reduced Emissions from Deforestation and Forest Degradation
RoC	Republic of Congo
RRI	Rights and Resources Initiative
SBTi	Science-Based Targets initiative
SD	Sustainable Development
SDG	Sustainable Development Goals
STAM	**S**trategies, Technocratic **T**ools, National **A**pproaches, and **M**ethods for IPLC and Other Stakeholder Engagement
TEK	Traditional Ecological Knowledge (TEK)
UNCCD	United Nations Convention to Combat Desertification
UNDRIP	United Nations Declaration on the Rights of Indigenous Peoples
UNDP	United Nations Development Program
UNEP	United Nations Environment Program
UNFCCC	United Nations Framework Convention on Climate Change

UNPFII	United Nations Permanent Forum on Indigenous Issues
USAID	United States Agency for International Development
VCM	Voluntary Carbon Market
WRI	World Resources Institute
WWF	Worldwide Fund for Nature

Preface

As a child, I was captivated by stamp collecting, which led me to explore distant places like the Fiji Islands, the Belgian Congo, and what was then Bechuanaland. My National Geographic subscription deepened this curiosity, prompting me to learn more about the world's new post-colonial nations, their political struggles, and the diversity of cultures. I began to wonder, with all the shiny new capital cities, new roads, and modern buildings, how would people living in traditional ways adapt to this change? Would the people I read about – who appeared from pictures to still be living "timelessly" in their cultures – eventually move to cities, leaving behind the very customs and practices that had shaped their lives for generations?

In the 1970s, I decided to go and physically experience those places and start getting some answers. I wasn't so much driven by academic goals – rather, I was mainly a traveler inspired by stamps and National Geographic. After a trip to Egypt and Sudan in 1979, I knew I wanted to work in Africa. What drew me in wasn't just the obscurity and size of the country or the diversity of people, wildlife, and landscapes. The energy of the place, the optimism and dreams, and the contrasts with the world outside Boston where I grew up were what led to me deciding that I wanted to see if in some small way I could help bridge this big chasm between the incredible histories and traditions that were still vibrant but beginning to change and the inevitability of this vague concept of modernization.

I actually tried to get a job on the Jonglei Canal project in Sudan, hoping to contribute to such an overwhelmingly challenging endeavor in one of the world's shest, swampy, malaria infested landscapes. Unfortunately, this led authorities to conclude I must have been a spy. They'd never met an American who apparently spoke Arabic well enough to get by, and who wanted to live in the swamps of southern Sudan.

In the early 1980s, I was hired as a socioeconomist on the Central Rangelands Development Project (CRDP) in Somalia. My job was to help the Somali government, USAID, the World Bank, and GTZ understand how Somali pastoral systems for managing rangelands and livestock worked and what, if anything, could (or should) be improved upon. The first phase of the $30 million project had been premised on the idea that Somali pastoralists were degrading the rangelands. After working closely with Somali pastoralists across all of central Somalia, in the regions where a decade later violent inter-clan warfare brought the country to the attention of the world stage, it became clear to me that Somali pastoralists were, in fact, managing the rangelands with incredible sophistication based on both traditional knowledge and the ability to live in the harshest of environments – and amazingly thrive. The range and livestock associations we formed with them were at most to tweak the inherent logic of what we understood they already were doing, one component of which was to avoid or, if necessary, manage conflict and competition over natural resources. Avoiding conflict through negotiating access agreements was key to the flexibility and resiliency of the Somali pastoral system.

This experience further reinforced my respect for Indigenous Peoples and Local Communities (IPLCs), and it has remained central to my interests and thinking ever since. Success in the Somalia project wasn't just about protecting the land or producing more livestock; it was fundamentally about understanding the culture, values, and practices – the motivations – of these truly charismatic people, who had been dismissed by the World Bank rankings at the time as the poorest people on earth. This dismissal occurred because the value of the herds they owned and continually sold at market was totally misunderstood and not accurately calculated. It was a pivotal moment that shaped the way I came to appreciate IPLC customary approaches to land use and their adaptations to dynamic geopolitical and subnational sociopolitical challenges. On top of that, how the rigors of the Horn of Africa's climate shaped these systems made the situation for me all the more compelling.

Fifty years later, that respect has only deepened, and this book is, in part, a reflection of those early experiences. The prevailing climate mitigation strategies promoted in the Global South today, whether through Nature-based Solutions (NbS), REDD+, or similar approaches, are often structured in ways that overlook the very elements I encountered while working alongside Somali pastoralists. These strategies too frequently ignore the complex and deeply embedded knowledge of IPLCs, applying external frameworks that are incompatible with local realities. The presumptuousness that local people do not understand that they are incapable of leading remains embedded in the culture of conservation and development, even as the rhetoric extolling traditional knowledge and local participation has become de rigueur for donors and all partners working not only in Africa but also across the entire Global South.

Whether it's the implementation of REDD+ projects, large-scale afforestation and reforestation (A/R) programs, or other NbS, the disconnect between the designers of these solutions and the people who are expected to implement them – the so-called co-beneficiaries – is glaring. By prioritizing carbon and its sequestration over the logic and imperatives of how and why people use the land the way they do, we preclude the possibility of achieving sustainability. This is exactly where I believe the strategies, tools, approaches, and methods (STAM) we are using – the system and the paradigm – are failing (see Box 0.2).

The rhetoric of inclusive, co-beneficial solutions in conservation and development has been compelling and persuasive to donors, thought leaders, and service providers embedded in the professional ecosystem. It has influenced both taxpayer donor programming and philanthropy for decades. But the reality is that much of what is designed fails to align with the lived realities of those whose land is being used for these projects (Conservation International, 2022). In fact, these projects are often driven by priorities that dismiss the very expertise and wisdom that local people hold, leading to not just failure but also often harm. This is why they too often fail when results and impacts are objectively measured by independent third parties, which is too often infrequent.

This book argues for a paradigm shift in the way we approach land-based climate mitigation in the Global South – one that not only shifts from technocratic, top-down methods but also fundamentally embraces the leadership and knowledge of IPLCs. That said, the entire burden cannot be put on the backs of IPLCs alone.

In my view, IPLCs must identify requirements for projects to work – if climate mitigation across vast swathes of the Global South is to contribute to the global common good. Moving forward, the future of successful, equitable climate mitigation in the Global south will lie in the willingness of those in positions of power to reflect and learn from them and from there, create space and opportunity to co-design, and negotiate strategies and terms. This is the only way sustainable solutions can be achieved. IPLCs must be tangibly and credibly empowered – not simply rhetorically – if sustainablity is to have a chance of being achieved.

Finally, viewed in a broader frame, the current wave of dramatic cuts to Overseas Development Assistance (ODA) offers a challenge and a chance to rethink the role of external aid in development and conservation. It opens the door to shift from strategies and approaches that inevitably engender dependency, toward more durable, collaboratively negotiated, climate mitigation models.

Paradoxically, these cuts in ODA financing for climate and bidiversity will necessitate greater tangible empowerment of IPLCs to meet climate, conservation, and sustainable development objectives. Now is the time to give IPLCs a real chance to show what they can do – not just for gaining greater security over their lands to increase equity – but as a key first step in climate mitigation, biodiversity conservation, central to the future of people and planet.

This book lays out a practical pathway for how that shift can begin in the land sector of the Global South.

Introduction

> Economic anthropologists see things differently. For them, the focus is the activities, relationships and systems through which objects are produced, circulate among people and ultimately are consumed, which take different forms in different societies and even in different parts of the same society. In this way, economic anthropology takes the rational market actors of conventional economic thought and places them in the world of people, relationships, systems, beliefs and values that begins with production and ends with consumption.
>
> – James Carrier, *Economic Anthropology* (2021)

Efforts to achieve climate mitigation in the land sector of the Global South (CMLSGS) are faltering (see Box 0.1). Despite extensive international investment, a cascade of strategies – including Nature-based Solutions (NbS), Natural Climate Solutions (NCS), Payment for Ecosystem Services (PES), and Agriculture, Forestry, and Other Land Use (AFOLU) – have consistently underperformed in delivering sustainable or just outcomes. These approaches often overlook what James Carrier describes as "the world of people, relationships, systems, beliefs, and values that begins with production and ends with consumption."

Box 0.1 CMLSGS Definition

Climate Mitigation in the Land Sector of the Global South (CMLSGS) refers to global climate policy addressing deforestation, biodiversity conservation, agriculture, and sustainable land-use practices. This book critically examines efforts by the international community – comprising donor agencies, governmental and intergovernmental institutions, and large international NGOs (BINGOs) – who consult with local civil society to design policies aimed at reducing deforestation and reversing unsustainable land-use patterns across the Global South. The central argument is that despite significant investment and rhetorical commitment, these externally driven policies and interventions repeatedly fall short due to fundamental mismatches with local contexts and limited genuine participation of Indigenous Peoples and Local Communities (IPLCs).

DOI: 10.4324/9781003380320-1

Economic anthropology offers a contrasting lens on economic decision-making – one that emphasizes how people assign value to resources, labor, and exchange through culturally grounded relationships, obligations, and social norms. From this viewpoint, decisions about land use and resource management are embedded in systems of meaning, not just profit maximization. Yet the international community – including governments, donors, and to some extent even Indigenous Peoples and Local Communities (IPLCs) – has approached CMLSGS, as a composite of interventions, largely through technocratic and market-based frameworks. Chapter 2 elaborates on why this orientation has weakened both the effectiveness and legitimacy of CMLSGS programs independent of the persistent technical challenges involving additionality, nonpermanence, and leakage.

Author's Note on Terminology

The acronym CMLSGS – Climate Mitigation in the Land Sector of the Global South – is introduced in this book as a unifying frame for analyzing a fragmented but influential set of international efforts. While no standardized term currently exists, CMLSGS enables systematic critique of land-use governance, REDD+, NbS, and other ecosystem-focused mitigation approaches aimed at reducing carbon emissions and promoting sustainability in the Global South. The term is used throughout the book to examine a domain that is operationally consequential but conceptually undertheorized.

This book applies an analytical framework grounded in substantive economic anthropology and political economy to explain why CMLSGS strategies have repeatedly underperformed – and what must change to achieve ecologically sound, socially just, and durable outcomes.

One of the interpretive lenses used throughout is economic anthropology, particularly its substantivist tradition. Rather than assuming that economic behavior is always rational and profit-maximizing, this perspective emphasizes how decisions are shaped by kinship, culture, moral obligation, and collective memory. In many IPLC communities, land and resource use is governed not by abstract models but by systems of social meaning and interdependence (Hann et al., 2024).

This helps explain why dominant climate mitigation strategies – designed by external actors using formalist economic assumptions that appear appropriate and realistic in the Global North – often fail to align with local incentives or governance realities. This is even before considering the gamut of intractable problems that have afflicted carbon offset market including nonadditionality, impermanence, leakage, double counting, perverse incentives, and the gameability of crediting systems (Romm et al., 2025).

The book proposes three interrelated design principles that climate mitigation strategies must meet to be viable for both people and planet in the Global South:

- **Appropriateness** – Do the strategies fit the local ecological, institutional, and cultural context?
- **Social Feasibility** – Are IPLCs and other stakeholders willing and able to engage because they see the strategies as fair, legitimate, and meaningful?
- **Technical Feasibility** – Are the strategies implementable given tenure systems, institutional capacity, and political constraints?

These are not merely evaluation tools; *they are design imperatives*. Throughout the chapters that follow, I argue that failure to meet even one of these conditions sharply reduces the likelihood of success. Many otherwise well-intentioned efforts break down precisely because they are conceived without this alignment.

These criteria are not simply evaluative; they are design principles. They should guide how the strategies, tools, approaches, and methods (STAM) behind mitigation efforts are conceived, adapted, and implemented. I argue that failure to meet even one of these conditions – appropriateness, social feasibility, or technical feasibility – substantially reduces the probability of success. These three design principles that are seemingly self-evident are rarely, if ever, explicitly written into STAM protocols.

The following chapters note that many well-intentioned initiatives have collapsed precisely because they were built on assumptions rather than stakeholder alignment. In doing so, they reproduce the governance failures they aim to solve to enable technical success. This recognition leads directly to the book's central argument: the need for a new theory of change grounded in negotiated governance and earned social license in CMLSGS.

These shortcomings are not marginal – they reflect deeper contradictions embedded in the prevailing CMLSGS paradigm: from the commodification of carbon to the persistent exclusion of IPLCs from decision-making authority. Through critique and case studies, this book highlights those systemic failures and offers an alternative model: one based on negotiated authority, social legitimacy, and locally appropriate pathways to climate mitigation.

The prevailing theory of change remains largely technocratic. It rests on the assumption that a carefully calibrated mix of economic incentives, baseline modeling, and MRV (Measurement, Reporting, and Verification) systems will curb deforestation and secure ecosystem services. This logic underpins REDD+ and many NbS programs. Yet despite extensive donor investment, the track record remains mixed at best, often marked by limited impact, poor durability, and contested legitimacy.

By contrast, this book proposes a theory of change that begins with IPLCs as co-equal designers of land-use strategies. Sustainability, in this view, must be socially negotiated and locally grounded, not dictated by external metrics or technocratic expertise. This approach starts with secure tenure, recognition of customary governance, and the alignment of incentives with IPLC worldviews and priorities shaped by lived experience.

This book calls for a paradigm shift rooted in methodological holism – an approach that embraces the complexity of real-world decision-making by drawing from both substantivist and formalist traditions in economic anthropology. Rather than privileging a single logic, it treats economic actions as "total social facts" (Carrier, 2021) – bundles of material practices, power dynamics, and symbolic meaning. In the context of climate mitigation, this framing enables strategies that reinforce IPLC priorities rather than displace them.

Such a paradigm shift centers IPLC agency, aligns incentives with culturally grounded governance systems, and redefines "participation" as co-equal decision-making enabled through negotiated governance. This is not just a normative ideal. It is a strategic necessity for designing STAM that are appropriate, feasible, and enduring across diverse Global South contexts.

The paradigm shift proposed here targets the STAM used by the international community, including intergovernmental organizations, donors, governments, BINGOs, and private actors. It calls for moving beyond top-down design and technocratic assumptions toward policy frameworks that are co-developed with IPLCs. In every national context, building social license – through trust, legitimacy, and shared authority – is the foundational step toward implementing credible, durable CMLSGS programs. Without this foundation, even seemingly technically sound initiatives are likely to fail.

Climate mitigation in the land sector cannot succeed through prioritizing carbon metrics and carbon marketing primarily. Lasting impact depends on whether STAM are contextually appropriate, socially legitimate, and technically feasible within real land use governance systems. Yet, as later chapters demonstrate, the prevailing paradigm has consistently failed on all three fronts. Climate finance remains distorted, carbon commodification dominates program logic, and technocratic delivery systems continue to sideline IPLCs – the very stewards most capable of sustaining ecological integrity over time. Prevailing land sector theory continues to promote the wrong assumptions about the problem and thus, the solutions.

STAM Must Be Politically and Institutionally Grounded

The dominant paradigm – shaped by intergovernmental agencies, donors, BINGOs, and increasingly private sector actors – remains anchored in a carbon-centric logic that prioritizes emissions targets and technical fixes. Market instruments like carbon credits, baseline modeling, and remote sensing have become central, often sidelining IPLC knowledge systems and privileging verification protocols over lived realities (African Arguments, 2024; IDS, 2024).

Over the past two decades, land-sector mitigation strategies in the Global South have revolved around avoided deforestation (REDD+), reforestation campaigns like the Great Green Wall, regenerative agriculture, and jurisdictional approaches such as J-REDD. While these efforts fall under the AFOLU umbrella, they are frequently plagued by misaligned incentives, technocratic reductionism, and limited local legitimacy.

Box 0.2 STAM Definition

STAM refers to the full suite of instruments and processes deployed to implement climate mitigation in the land sector of the Global South (CMLSGS). This includes the overarching *strategies* (such as REDD+, NbS, PES, or AFOLU); the *tools* used to measure, deliver, and monitor them; the *approaches* adapted at national and subnational levels; and the *methods* through which stakeholders are engaged – often ineffectively. The book critiques prevailing STAM for being technocratic, top-down, and frequently misaligned with social, institutional, and ecological realities on the ground.

Climate science has been essential in framing the global urgency of climate action. But urgency alone is not a strategy. In the Global South, "science-based targets" are frequently operationalized through carbon modeling and digital MRV systems that lack grounding in the political and social realities of land governance. This creates a persistent mismatch between scientific ambition and implementation capacity – one that undermines both credibility and practical outcomes.

This narrow framing around carbon management not only distorts priorities; it also places disproportionate burdens on IPLCs. Communities are often displaced or restricted in their land use, with agriculture and customary systems of stewardship sidelined by externally imposed models (Modelling Injustice, 2024). Projects are too often launched before feasibility or local appropriateness is properly assessed, leading to dispossession, poor design, and escalating forms of "green grabbing" (IDS, 2024).

While many professionals involved in CMLSGS projects are committed and well-intentioned, the level of community pushback is rising. This resistance is not a failure to cooperate – it is a rational response to flawed assumptions about governance, tenure, and trust. When mitigation policies fail to demonstrate social feasibility and appropriateness, they will not earn the social license needed for legitimacy or long-term success.

The consequences are increasingly visible: displaced livelihoods, erosion of customary authority, and data-driven interventions that prioritize metrics over customary rights. Projects that disregard trust and tenure are not just technically weak – they are socially infeasible and thus, unsustainable. Resistance, in this context, signals the failure of imposed and ineffective design, not of community commitment for genuine solutions.

Why a New Theory of Change Is Needed

To address the shortcomings of the prevailing paradigm and its embedded STAM, this book proposes a new theory of change for CMLSGS – one grounded in earned social license, negotiated governance, and a definition of sustainability that goes beyond carbon to include equity, legitimacy, and long-term stewardship.

The dominant theory of change has been largely technocratic. It assumes that the right mix of economic incentives, emissions baselines, and MRV systems will reliably deliver avoided deforestation and ecosystem service benefits. Although this logic underpins REDD+ and many NbS programs, the empirical results remain underwhelming, even after years of donor investment and technical refinement.

By contrast, the alternative theory of change advanced here begins with social license – the trust, legitimacy, and consent required for climate interventions to be accepted and sustained by IPLCs. Without social license, or a "social contract", even well-designed programs fail to gain traction. It is not a procedural formality. Rather, it is the foundational condition for credibility, feasibility, and scale.

Building on that foundation, this book introduces negotiated governance as a structural solution. The conceptual model strives to position IPLCs, governments, and the private sector as co-developers of the STAM required to achieve CMLSGS, with other stakeholders including donors, BINGOs, etc as technical support to that core. Through structured, iterative, and transparent dialogue, stakeholders co-develop land-use strategies that reflect shared priorities and mutual accountability. Unlike consultation, often treated as a post hoc procedural step, negotiated governance begins with shared authority and consent from the outset. This is a strategy leading to a social contract.

Under negotiated governance, IPLCs are not passive recipients or token "co-beneficiaries." They are architects of strategy – helping define goals, shape incentives, and monitor implementation in ways that are context-specific, durable, and grounded in customary governance systems.

This new theory of change rests on two interlocking premises:

1. **Sustainability must be understood holistically, not reduced to carbon metrics alone.** It includes the social, ecological, and economic dimensions of long-term land stewardship.
2. **Incentives must reward stewardship, not commodified carbon.** Effective mitigation must treat land use as an integrated social–ecological system, not a set of isolated carbon components. Systems that reinforce reciprocal land care, community-based monitoring, and customary governance are far more likely to earn legitimacy and deliver lasting impact. Leach and Fairhead (1996) clearly demonstrated in their study of West African forest–savanna mosaics, that local communities have historically enriched – not degraded – their landscapes through embedded land-use systems shaped by social norms, ecological knowledge, and intergenerational responsibilities. While their work predates the climate mitigation agenda, it powerfully illustrates that such systems, common across the Global South, are not obstacles to climate action – they are its foundation.

This shift requires a rethinking of not only the technical tools but also the broader governance architecture that underpins CMLSGS. It calls for moving away from top-down delivery and toward Negotiated Governance Platforms (NGPs), and from box-checking participation to shared control over climate decision-making. NGPs are typically tripartite in form, engaging the three core actors in climate governance and finance: governments, IPLCs, and private-sector entities, including impact investors and philanthropic funders. However, they are not limited to these three; they may also include multilateral institutions, local authorities, or civil society actors where appropriate. Crucially, NGPs are not designed to override embedded IPLC governance systems; they are meant to align finance, regulatory frameworks, and external mechanisms with those systems without engendering displacement.

Until now, implementation and funding flows have been largely shaped by international intermediaries, such as multilateral agencies, Big Conservation (and Development) NGOs, and consulting firms, whose dominance has failed to produce meaningful results at either the biophysical or human development level. Their underperformance, particularly in delivering on both carbon and SDG outcomes, is the clearest rationale for a fundamental paradigm shift.

The problem is not merely one of poor implementation. It is conceptual. For two decades, mitigation efforts have relied on market tools and science-framed urgency as substitutes for political negotiation, institutional trust, and governance reform. These approaches often misread power relations, disregard customary tenure, and impose standardized incentives in contexts where long-term stewardship depends on relationships, not metrics.

The theory of change advanced in this book reverses those assumptions. It begins by recognizing IPLCs as co-governors of land and ecosystems, not as peripheral actors or conditional beneficiaries. Mitigation must become a negotiated process, with strategies co-developed through inclusive governance, shared authority, and a grounded understanding of place-based realities.

This requires more than symbolic inclusion. It means shifting from extractive project cycles to institutional transformation – from short-term outputs to long-term partnerships built on legitimacy and accountability. When institutionally supported and adequately resourced, IPLC-led stewardship consistently outperforms externally managed systems in protecting ecosystems, reducing emissions, and sustaining biodiversity (Garnett et al., 2018; IUCN, 2022; RRI, 2023).

To operationalize this transformation, the book proposes a layered framework: beginning with capacity building and legal empowerment, progressing through participatory scenario analysis and joint decision-making, and culminating in institutional systems that embed IPLC leadership, secure tenure, and long-term rights.

In this model, social license is not a checkbox – it is the condition for legitimacy, scale, and durability. Without it, even technically sound programs will falter. With it, plural, locally grounded strategies can emerge that reflect the lived priorities, governance logic, and cultural systems of IPLCs.

This theory of change challenges the assumptions that have guided land-sector mitigation for two decades. It replaces the fantasy of scalable technocratic fixes with a politics of negotiated alignment. It affirms that legitimacy is not ancillary to effectiveness – it is its precondition.

The Conceptual Framework Guiding This Book

The conceptual framework used throughout this book draws on insights from economic anthropology, institutional theory, and Indigenous governance systems (see Figure 0.1). It is grounded in decades of public-domain reporting and field evidence

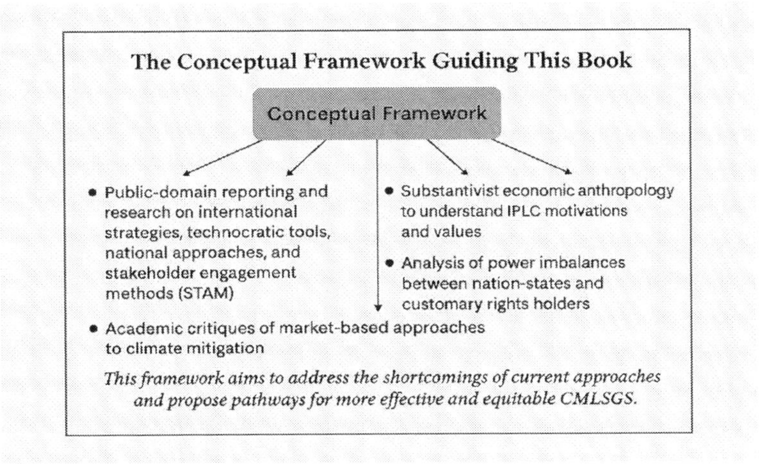

Figure 0.1 The Conceptual Framework Guiding This Book

from international development, biodiversity conservation, REDD+, agroforestry, and sustainable agriculture.

At its core is adherence to the principles that STAM must be evaluated not only for technical merit but also for contextual appropriateness and feasibility. This means assessing how global strategies play out in specific national, subnational, and community-level settings – settings where political economy, tenure systems, and customary norms are often as or more determinative than carbon baselines.

A substantivist lens from economic anthropology helps reveal the disconnect between dominant models, such as carbon markets and PES schemes, that assume rational, incentive-based behavior, and the relational, collective, and often non-monetary ways IPLCs interact with land and resources. When external actors presume that financial incentives alone, or primarily, will drive behavioral change, they overlook how decisions are actually shaped by customary authority, cultural meaning, and multigenerational planning horizons.

This framework also brings into focus the persistent power imbalances between state institutions and customary rights holders. Genuine reform requires more than policy tweaks: It demands a transformation in how tenure is recognized, how climate finance is accessed, and how mechanisms like REDD+ and NbS are governed, validated, and held accountable.

Throughout the chapters that follow, case studies and critiques illustrate how dominant systems continue to marginalize IPLCs, even as growing evidence shows that IPLC leadership offers the most credible and sustainable pathway to effective mitigation.

This book does not present a single blueprint. Instead, it lays the foundation for developing NGPs – institutional models that recognize IPLCs as co-equal decision-makers and create structured processes for aligning climate strategies with shared goals of sustainability, equity, and legitimacy.

IPLCs as Central Actors in Climate Mitigation

IPLCs manage or influence land use across up to 50% of the world's terrestrial surface (Rights and Resources Initiative, 2023). They steward 80% of global biodiversity hotspots and hold significant portions of the world's remaining intact forests and carbon sinks (FAO, 2025; WRI, 2025). Yet despite this centrality, they remain marginalized in most climate mitigation policies and programs (FAO & FILAC, 2021).

The evidence is compelling: IPLC-managed lands consistently exhibit lower deforestation rates, greater ecological resilience, and more sustainable land use than state- or corporate-controlled areas (Dooley et al., 2022; Forest Trends, 2023). In the Amazon, for example, Indigenous-led conservation sequesters 55% more carbon per hectare than state-managed forests (Dooley et al., 2022). These outcomes are not incidental. They reflect systems rooted in customary governance, traditional ecological knowledge (TEK), and long-term relationships to land.

Nonetheless, prevailing mitigation frameworks continue to treat IPLCs as peripheral actors – at best, "co-beneficiaries," not co-decision-makers. This structural exclusion has practical consequences. Insecure tenure, lack of recognition for customary rights, and exclusion from governance processes leave IPLCs vulnerable to land grabs, displacement, and externally imposed agendas (FAO, 2025). Economic barriers – such as the high cost of formal land titles and the complexity of compliance systems – further entrench exclusion, even when programs claim to be "inclusive."

Too often, IPLC engagement is reduced to stylized consultation or procedural Free, Prior, and Informed Consent (FPIC) processes that fall far short of transferring real power (Accountability Framework Initiative, 2020). These tokenistic approaches misread what IPLCs bring to the table. When genuinely empowered, IPLCs are not merely "capable" – they are among the most consistent and credible stewards of land and biodiversity available to the global climate community (Artelle et al., 2019).

Recognizing IPLCs as lead negotiators – rather than consulted stakeholders – is not just a matter of justice. It is a matter of feasibility. Lasting outcomes in land-based climate mitigation do not depend on the strength of carbon models or the size of donor pledges, but on whether governance systems are grounded in legitimacy, capability, and shared authority. The chapters that follow examine how the prevailing system has failed on all three fronts – and how a shift toward negotiated governance and co-equal IPLC leadership offers a more credible path forward.

Forging Social License and Negotiated Platforms

Credible, lasting CMLSGS requires more than good design or new technologies. It requires legitimacy. That legitimacy, what many call social license, cannot be assumed or imposed. It must be earned through trust, transparency, and meaningful participation (Gunningham et al., 2004).

In this context, social license means more than community acceptance. It reflects a shared understanding among IPLCs, governments, and external actors about the purpose, rules, and outcomes of a given intervention, including how risks are managed and how benefits are distributed. Without this foundation, even well-intentioned projects often fail to take root.

Earning social license begins with recognizing IPLCs as rights-holders and governance partners, not as stakeholders to be consulted after decisions are made. It also requires rejecting the checkbox approach to participation. When "inclusion" is reduced to late-stage consultation or symbolic consent, communities are not engaged – they are alienated.

That is why this book proposes negotiated governance as the operational foundation for climate mitigation. Negotiated governance is not extended consultation – it is shared authority. It creates structured platforms where IPLCs, governments, and private or philanthropic actors jointly determine land-use priorities, performance expectations, and long-term stewardship agreements.

In this model, FPIC is not a box to tick at the beginning – it is an ongoing commitment. It centers relationship-building, co-development, and the use of dispute resolution mechanisms that communities view as legitimate, not just those recognized externally.

Several entrenched features of current mitigation efforts make social license hard to secure:

- Top-down programming that ignores local priorities
- Technocratic tools that prioritize metrics over meaning
- Rigid performance targets that leave no room for adaptation
- Opaque financing and benefit-sharing, which undermines trust

Reversing this pattern requires more than better-designed projects. It requires institutional humility. Donors, governments, and implementers must accept that they cannot lead from above. Their role is to support the enabling conditions – secure tenure, accessible information, fair financing, and negotiation capacity – that allow IPLCs to lead from within.

Only then can social license support climate mitigation that is scalable, durable, and just.

The Problem With Carbon Commodification

One of the most entrenched features of the prevailing paradigm is the commodification of carbon – the belief that greenhouse gas reductions can be isolated, quantified, packaged, and traded across jurisdictions. This approach converts complex ecological functions, such as forest integrity, biodiversity, and carbon sequestration, into standardized units of exchange. These units are then embedded in markets that prioritize tradable value over governance context, social legitimacy, or local priorities.

This logic underpins REDD+, voluntary carbon markets, nature-based offsets, and many PES schemes – all marketed as "win-wins" for people and planet.

These carbon-centric models often displace IPLCs, undermine customary tenure systems, and deliver questionable environmental benefits (Bayrak & Marafa, 2016; Beymer-Farris & Bassett, 2012; Fletcher & Breitling, 2012). Rather than addressing structural barriers like insecure land rights or weak governance, these models impose narrow technical fixes that are often misaligned with local realities.

Large-scale initiatives such as the Great Green Wall have faltered because they were launched without local legitimacy, relied on weak or bypassed governance, and ignored the incentive systems that sustain long-term land care (Modelling Injustice, 2024). Even more sophisticated efforts like jurisdictional REDD+ have struggled to correct asymmetries in power, often reinforcing the very hierarchies they were meant to reform (Proforest Initiative, 2023). Relying primarily on state institutions to establish land use rules for local natural resource management systems are destined to fail.

These failures are not just design flaws. They reflect a fundamental misreading of the problem. Climate mitigation cannot be solved by exporting carbon reductions alone. It requires transforming how land is governed and how success is defined (African Arguments, 2024).

Until the underlying logic of commodified carbon is replaced with an ethic of embedded stewardship, climate programs will continue to face resistance, inequity, and underperformance – especially in IPLC territories, where the social and ecological costs of failure are most acutely felt.

Aligning Incentives With IPLC Values and Structural Realities

For land-based climate mitigation to succeed, it must reflect how people actually make decisions, not how economic models assume they do. In IPLC contexts, decisions about land use and stewardship are shaped by a complex web of social norms, customary institutions, kinship obligations, spiritual relationships to place, and structural constraints imposed by state and global actors. While market logic may be present, it is rarely the primary driver.

Yet dominant mitigation strategies – such as NbS, NCS, and PES – often treat behavior as linear and transactional, assuming that financial incentives or market access will produce uniform, rational responses. This reflects a formalist bias rooted in neoclassical and neoliberal economics, which centers on individual self-interest and optimization while ignoring deeper structures of meaning, authority, and community. Even when donors conduct political economy analyses, these rarely extend beyond elite actors or formal state institutions.

Economic anthropology, particularly the substantivist tradition, offers a more appropriate lens. It views economies as socially embedded and historically situated. As Wilk and Cliggett (2007) explain, even in so-called non-market societies, economic choices are shaped as much by moral obligation and collective well-being as by material rationality. This insight – long familiar to anthropologists and community-based practitioners – has been largely ignored in the design and implementation of REDD+, and more recently, NbS.

Despite increasing rhetorical support for IPLC rights, the core assumptions underlying mainstream mitigation strategies continue to reflect market-centric models that are only partially applicable to the lived realities of IPLCs across the Global South where land use systems are evolving from customary to hybrid forms. Blanton et al. (2024) provide a partial overview of issues and challenges market models face in Latin America working with IPLCs.

TEK, also commonly referred to as Indigenous Knowledge (IK), which is more widely embraced and conceptually broader, illustrates the frequent disconnect. Indigenous stewardship often prioritizes reciprocity, restraint, and long-term ecosystem balance over short-term maximization. Studies from Latin America, for example, show that Indigenous-managed forests sustain higher biodiversity and more stable carbon stocks than state or private-sector equivalents (Dalton, 1961; Weichenrieder, 2024).

Climate strategies that ignore these dynamics risk not only failure but also active backlash. When external incentives override local governance or treat communities as cost–benefit calculators, the result is often mistrust, disengagement, and the erosion of the very stewardship on which mitigation depends.

Embedding incentives in local institutions – and aligning them with IPLC worldviews – is not a compromise. It is a prerequisite for success.

Toward a Layered Strategy for Transformational Change

Supporting IPLC stewardship is not a one-off intervention. It requires coordinated action at multiple levels – starting with foundational support and moving toward deeper reforms that shift power, reshape incentives, and align institutions with IPLC leadership.

This book proposes a layered strategy for transformational change, grounded in negotiated governance and built on the premise that IPLC leadership is essential not just to justice but also to the feasibility, credibility, and long-term viability of climate mitigation.

Level 1 – Foundational Support Within the Current Paradigm

Despite growing recognition, many IPLCs remain structurally excluded from climate programs due to insecure tenure, limited legal recognition, cultural biases, and high transaction costs. Addressing these challenges requires:

- Capacity Building: Legal literacy, negotiation training, and support for participatory governance (e.g., IFAD-backed mapping initiatives) can strengthen tenure and community leadership (IFAD, 2022).
- Financial Access: Simplified grant mechanisms and direct-to-community funding reduce dependency on intermediaries. The $1.7 billion Forest Tenure Pledge is a step in this direction (Sutherland, 2021).
- Legal Advocacy: Implementing FPIC and strengthening adherence to instruments like the United Nations Declaration on the Rights of Indigenous Peoples (United Nations, 2007) is essential to building trust and legitimacy.

Level 2 – Developing Analytical and Negotiation Capacity

IPLCs need more than inclusion – they need tools to analyze trade-offs, develop strategies, and negotiate on equal footing. As Chapter 7 explores, this includes scenario planning, cost-benefit analysis, and co-development of long-term land-use visions.

One early model was the Community Options Analysis and Investment Toolkit (COAIT), developed under USAID funding in the early 2000s. While no longer in use, COAIT demonstrated how participatory cost-benefit and risk analysis, mapping and, spatial analysis, and ecological scenario planning could empower IPLCs to define land-use priorities from the ground up.

Level 3 – Aligning Interventions With IPLC Worldviews and Values

Climate programs must do more than simply consult – they must account for IPLC institutions, governance norms, and moral economies. Substantivist economic anthropology explains why land-use decisions are embedded in systems of meaning, kinship, and intergenerational responsibility – not just market logic. It is fit for purpose to assist in achieving CMLSG.

One striking example of what is needed comes from New Zealand, where Māori communities practicing *kaitiakitanga* – a customary ethic of guardianship – achieved 32% higher reforestation outcomes than comparable state-managed areas (CIF Report, 2012). While not from the Global South, the lesson is clear: When IK is genuinely integrated, sustainability improves. Substantivist economic anthropology offers a framework that can help.

IPLCs are not a panacea for global climate failure. In many contexts, they lack legal standing, institutional access, or technical tools to fulfill the roles increasingly assigned to them. At the same time, IPLCs and IK are often celebrated rhetorically in Global North policy – but rarely matched by investments in governance reform, direct finance, or long-term capacity building.

Yet across much of the Global South, IPLCs represent the hinge point between success and failure. Without their leadership, climate mitigation, biodiversity conservation, and rural development will continue to underperform. Acknowledging both their transformative potential and structural constraints is essential for designing strategies that are not just aspirational but also viable.

A Holistic Framework for IPLC Negotiation and Decision-Making

To reframe IPLCs as negotiators – not beneficiaries – requires more than participation in pre-designed projects. It means shifting decision-making authority and designing governance platforms that reflect how communities actually make land-use choices.

This book draws on debates in economic anthropology to explain why this shift is essential. In particular, it contrasts two longstanding perspectives:

- Substantivist anthropologists like Karl Polanyi have long argued that economic behavior is embedded in social institutions – not driven by abstract market rationality (Polanyi et al., 1957). People act not simply to maximize profit but in accordance with values, obligations, and cultural norms. In many Indigenous and peasant societies, land is not a commodity – it is a relational space, governed by reciprocity, redistribution, and stewardship.
- Formalist anthropologists, by contrast, assert that rational, cost–benefit thinking applies universally, across cultures and systems. While formalist insights are useful for modeling behavior, they often fail to capture the meaning systems and governance structures that drive real-world decision-making in IPLC contexts.

Other schools – Marxist, structuralist, and feminist – have expanded this foundation by highlighting how global markets, donor policies, elite capture, and historical inequality shape what options are available to IPLCs in the first place.

Rather than endorsing a single tradition, this book adopts a methodological holism, treating land-use decisions as "total social facts" (Carrier, 2021). These are decisions shaped simultaneously by material needs, political relationships, cultural meanings, and institutional power.

This lens allows a more honest and accurate approach to STAM design. It demands asking not only whether a tool works in theory but also whether it is appropriate, feasible, and just – in context. It also helps explain why simplified FPIC procedures or market-based incentives often fail: They overlook the deeper forces – social, symbolic, and structural – that drive real-world choices.

The message of this book is clear: Without IPLC leadership, land-based climate mitigation in the Global South will continue to falter – technically, socially, and politically. A new model is urgently needed – one that replaces commodified carbon logic with embedded stewardship, prioritizes appropriateness and feasibility, and builds legitimacy through negotiated governance. The chapters that follow unpack this model through critique, evidence, and alternative proposals. An effort is made to show how a new generation of climate strategies can move from exclusion and failure toward equity, credibility, and lasting impact.

Positioning the Framework for What Follows

The conceptual framework guiding this book is practical, not idealistic. It is grounded in the realities of policy implementation and the institutional, social, and economic conditions that shape success or failure.

Throughout the chapters that follow, this book examines the persistent gap between evidence and action, asking not only why programs have failed but also how structural change can be achieved through context-specific, negotiated solutions.

- Chapter 1 examines why the disconnect between knowledge and implementation persists across land-based mitigation efforts.
- Chapter 2 explores how political polarization and fragmented institutions block reform.
- Chapters 3–6 assess the dominant tools and strategies: NbS, carbon markets, finance architectures, and stakeholder engagement models.
- Chapter 7 outlines a forward-looking governance model based on social feasibility, negotiated authority, and institutional transformation.

A Personal Anecdote Why Communities Are Key

That experience began for me in the early 1970s, first with the pastoral Samburu and Maasai in Kenya, and then with the Bambuti in the Ituri Forest – one of the most ecologically rich regions of the Congo Basin rainforest. What all of these communities shared was a depth of TEK that was, quite simply, off the charts. Their systems of land and resource stewardship were highly adapted, intuitive, and grounded in relationships that spanned generations.

A few years later, I became the project socioeconomist for what was then the largest rangeland management initiative ever funded in Africa: the Central Rangelands Development Project in Somalia. At the time, Somalia was labeled by the World Bank as one of the poorest countries in the world. But the pastoral families I worked with across central Somalia owned, on average, $130,000 worth of livestock – when delivered to port in Jeddah for the Hajj. Surface appearances aside, this wasn't poverty. It was a dynamic economic system that worked, built around inter-clan recognition of customary grazing zones and a sophisticated set of negotiated agreements known as *Heer*. These customary contracts – oral, flexible, and enforced by clan elders – established reciprocal rights and obligations around access to grazing lands, water points, and other critical resources. Embedded within the broader Somali legal tradition of *Xeer*, these agreements enabled communities to manage drought risk, reduce conflict, and maintain adaptive mobility. International agencies saw degradation; what they missed was a functioning governance system rooted in negotiation, reciprocity, and resilience.

That misunderstanding between perceived failure and actual functionality sits at the heart of this book. Many customary systems still function, or could function effectively, if given the chance, with secure tenure, stronger IPLC capacity, and genuine space for communities to articulate not only their needs but also their capabilities and priorities. Too often, those systems are bypassed in favor of technocratic models that emphasize carbon metrics while undermining the very conditions that sustain ecosystems, cultures, and livelihoods. What follows is an effort to re-center those systems, not by romanticizing them, but by recognizing their logic, appreciating their potential, identifying where they can benefit from strengthening or adjustment, and creating space for IPLCs to make the case and negotiate the terms for how land is to be governed. Only then can land-based climate mitigation be appropriate, feasible, and grounded in the people most invested in its success.

Finally, this book is not anti-science. It is opposed to *simplistic, top-down technocracy* that treats complexity, or any resistance to the carbon market-driven paradigm, as friction to be engineered away. It is in favor of credible, evidence-based science, not the recycled orthodoxies that prevail even when challenged by field experience, empirical data, and community knowledge. It opposes simplistic assumptions that financial incentives primarily will shift community behaviors, failing to account for how decisions are shaped by customary authority, cultural meaning, and multigenerational planning horizons.

The term STAM is used throughout the book to refer to the operational instruments used to implement land-based mitigation. These cannot be evaluated on technical merit alone. STAM must be assessed against three interlocking conditions: appropriateness to the local context, feasibility within existing institutional and governance realities, and legitimacy as judged by IPLC systems and norms. When these criteria are not met – for example, when STAM is imported wholesale without fine-tuned adaptation – projects falter, and trust erodes. These conditions are not just evaluation tools; they are design principles.

Technocratic models, especially those built around remote sensing, carbon baselines, and MRV systems, often fail these tests. They overemphasize measurable biophysical variables and downplay political complexity. The result leads to a form of climate governance that is neither credible nor sustainable in IPLC territories. This book argues for negotiated governance that centers IPLC authority, respects plural legal systems, and embeds STAM in context-driven processes.

The chapters that follow examine how this can be done – starting with why the current paradigm has failed to deliver on its promises.

References

Accountability Framework Initiative. (2020). Operational guidance on free, prior and informed consent: Requirements, best practices, and practical considerations for companies to fulfill their obligation to secure the FPIC of Indigenous Peoples and local communities (Minor revision: May 2020). Gland, Switzerland. https://accountability-framework.org/fileadmin/uploads/afi/Documents/Operational_Guidance/OG_FPIC-2020-5.pdf

African Arguments. (2024). Modelling injustice: South experts call for climate model paradigm shift. https://africanarguments.org/2024/04/modelling-injustice-south-experts-call-for-climate-models-paradigm-shift/

Artelle, K. A., Zurba, M., Bhattacharyya, J., et al. (2019). Supporting resurgent Indigenous-led governance: A nascent mechanism for just and effective conservation. *Biological Conservation*, 240, 108284.

Bayrak, M. M., & Marafa, L. M. (2016). Ten years of REDD+: A critical review of the impact of REDD+ on forest-dependent communities. *Sustainability*, 8(6), 620.

Beymer-Farris, B. A., & Bassett, T. J. (2012). The REDD menace: Resurgent protectionism in Tanzania's mangrove forests. *Global Environmental Change*, 22(2), 332–341.

Carrier, J. G. (2021). *Economic Anthropology*. Newcastle upon Tyne: Agenda Publishing.

CIF Report. (2012). *The Contribution of Traditional Knowledge and Technology*. Climate Investment Funds.

Conservation International. (2022). *Indigenous Negotiations Resource Guide*. Retrieved October 4, 2025, from https://www.conservation.org/projects/indigenous-negotiations-resource-guide

Dalton, G. (1961). Economic theory and primitive society. *American Anthropologist*, 63(1), 1–25.

Dooley, K., Keith, H., Larson, A., Catacora-Vargas, G., Carton, W., Christiansen, K.L., Enokenwa Baa, O., Frechette, A., Hugh, S., Ivetic, N., Lim, L.C., Lund, J.F., Luqman, M., Mackey, B., Monterroso, I., Ojha, H., Perfecto, I., Riamit, K., Robiou du Pont, Y., & Young, V., 2022. *The Land Gap Report 2022*. Available at: https://www.landgap.org/

FAO. (2025). *Collective Tenure Rights and Climate Action in Sub-Saharan Africa*. Rome: Food and Agriculture Organization of the United Nations.

FAO & FILAC. (2021). *Forest Governance by Indigenous and Tribal Peoples*. Rome: FAO.

Fletcher, R., & Breitling, J. (2012). Market mechanism or subsidy in disguise? Governing payment for environmental services in Costa Rica. *Geoforum*, 43(3), 507–517.

Forest Trends. (2023). *Forest Trends Impact Report 2023: The Magic of Coalitions*. https://www.forest-trends.org/publications/forest-trends-impact-report-2023/

Garnett, S. T., Burgess, N. D., Fa, J. E., Fernández-Llamazares, Á., Molnár, Z., Robinson, C. J., . . . & Watson, J. E. M. (2018). A spatial overview of the global importance of Indigenous lands for conservation. *Nature Sustainability*, 1(7), 369–374.

Gunningham, N., Kagan, R. A., & Thornton, D. (2004). Social license and environmental protection: Why businesses go beyond compliance. *Law & Social Inquiry*, 29(2), 307–341.

Hann, C., Hart, K., & Wilk, R. (2024). Revisiting the formalist-substantivist debate in economic anthropology. *Anthropology Review*, 7(1), 55–82.

IFAD. (2022). *Indigenous Peoples' Collective Rights to Lands, Territories and Natural Resources*. Rome: International Fund for Agricultural Development.

Institute of Development Studies. (2024). Why policy makers need to stop treating climate change in isolation. *IDS Opinions*. https://www.ids.ac.uk/opinions/why-policy-makers-need-to-stop-treating-climate-change-in-isolation/

IUCN. (2022). Inclusive conservation initiative. https://www.iucn.org/

Leach, M., & Fairhead, J. (1996). *Misreading the African Landscape: Society and Ecology in a Forest-Savanna Mosaic*. Cambridge: Cambridge University Press.

Polanyi, K., Arensberg, C. M., & Pearson, H. W. (1957). *Trade and Market in the Early Empires: Economies in History and Theory*. Glencoe, IL: Free Press.

Proforest Initiative. (2023). *Respecting rights of Indigenous Peoples and local communities in landscape initiatives*. Retrieved October 4, 2025, from https://www.proforest.net/

Rights and Resources Initiative (RRI). (2023). *2022 Annual Report*. Washington, DC: Rights and Resources Initiative. Retrieved April 17, 2025, from https://rightsandresources.org/wp-content/uploads/Annual-Report-2022-v6-WEB.pdf

Romm, J., Lezak, S., & Alshamsi, A. (2025). Are Carbon Offsets Fixable? *Annual Review of Environment and Resources*, 50, 649–680. https://doi.org/10.1146/annurev-environ-112823-064813

Sutherland, L. (2021, November 2). $1.7 billion pledged in support of Indigenous and local communities' land tenure. *Mongabay News*. https://news.mongabay.com/2021/11/1-7-billion-pledged-in-support-of-indigenous-and-local-communities-land-tenure/

United Nations. (2007). United Nations Declaration on the Rights of Indigenous Peoples. *United Nations Office of the High Commissioner for Human Rights*. https://www.ohchr.org/en/indigenous-peoples/un-declaration-rights-indigenous-peoples

Weichenrieder, S. (2024, August 26). Leveraging Indigenous knowledge for effective nature-based solutions in the Arctic. *The Arctic Institute's Arctic Extractivism Series 2024*. The Arctic Institute. https://www.thearcticinstitute.org/leveraging-indigenous-knowledge-effective-nature-based-solutions-arctic/

Wilk, R. R., & Cliggett, L. (2007). *Economies and Cultures: Foundations of Economic Anthropology* (2nd ed.). Boulder, CO: Westview Press.

WRI. (2025). 6 Strategies that achieve climate mitigation and adaptation simultaneously. *World Resources Institute*. https://www.wri.org/insights/strategies-achieve-climate-mitigation-adaptation-simultaneously

1 Disconnects Between Evidence and Action in Land-Sector Climate Mitigation

As the Introduction argued, the dominant mitigation paradigm – rooted in carbon metrics and technocratic delivery – often bypasses the political, cultural, and institutional dynamics that determine success. This chapter explores why that disconnect persists despite overwhelming evidence of its costs.

This book focuses on Climate Mitigation in the Land Sector of the Global South (CMLSGS), particularly on strategies aimed at avoiding emissions and carbon removals (Wang et al., 2023). Pathways for integrating NbS into national climate strategies have been extensively addressed in UNDP technical guidance for governments preparing or updating their NDCs (United Nations Development Programme, 2019).

While rooted in a global scientific consensus on the urgency of climate action (Fawzy et al., 2020), these efforts are often implemented through tools and engagement models that fail to align with the development priorities, governance systems, and cultural realities of Indigenous Peoples and Local Communities (IPLCs) (IISD, 2024). Annual assessments by the United Nations Environment Programme consistently show that current trajectories remain far off track from targets, with the emissions gap widening despite rising rhetoric and financing efforts (UNEP, 2021, 2022).

The disconnect is not merely technical or financial – it reflects deeper justice concerns, including the Global North's historical emissions debt and contested responsibilities for mitigation, which have been central to international climate negotiations under the Paris Agreement (Chancel, 2021; Hickel, 2020; Humpenö der et al., 2022; Lefstad & Paavola, 2023; UNFCCC, 2015).

The dominant paradigm draws legitimacy from climate science, especially the IPCC's synthesis of over 97% scientific consensus on the causes and urgency of global warming (International Energy Agency, 2021; IPCC, 2019, 2021, 2022b). In particular, a recent global synthesis documents the continued escalation of extreme heat, directly attributing these trends to anthropogenic climate change (Climate Central et al., 2024). Yet this consensus has not translated into operational clarity, particularly in land-based sectors where tenure insecurity, weak institutions, and contested land rights make "rapid scaling" inherently risky. Expert surveys indicate a growing uncertainty regarding the effectiveness and risks associated with NbS (Baum et al., 2024). Additional analyses highlight

DOI: 10.4324/9781003380320-2

that land-based mitigation strategies in IPCC pathways could exacerbate food insecurity in low- and middle-income countries, posing serious equity risks (Jaiswal et al., 2024). At the same time, new research reinforces that the macroeconomic damages of climate change may be far greater than previously estimated, heightening pressure for mitigation (Bilal & Känzig, 2024; Newman & Noy, 2023). In contrast, critics like Lomborg (2020) argue that such mitigation efforts risk harming the poor more than they help.

In practice, this scientific urgency is often operationalized through interventions like REDD+, NbS (IUCN, 2019), and carbon offset programs. These models aim to produce quantifiable emissions reductions but frequently ignore governance fragility, elite capture, and IPLC exclusion – factors that determine whether interventions are viable and just in practice (Dawson et al., 2021; Haya et al., 2023). Measurement, Reporting, and Verification (MRV) systems, for example, rely on satellite-derived baselines while bypassing local tenure systems and unresolved land conflicts. These tools may appear scalable, but they are highly technocratic, and often socially unfit and institutionally ungrounded.

The issue is not with science itself, but with how scientific urgency is translated into policy, often as justification for deploying standardized tools and metrics without regard for grounded governance systems or local authority. This mismatch between epistemic authority and operational reality has created a widening credibility gap in land-sector mitigation across the Global South.

That credibility gap is most visibly reflected in the climate finance system. Governments and civil society leaders from the Global South have consistently demanded more ambitious, reparative financing, emphasizing that climate justice requires recognition of historical emissions, differentiated responsibilities, and the burdens disproportionately falling on the least responsible (Chancel, 2021; Hickel, 2020). At COP28 and in subsequent multilateral forums, coalitions including the Climate Action Network and the Global Campaign to Demand Climate Justice reiterated calls for at least $5 trillion per year in climate finance as a form of reparations (Climate Action Network, 2024; Demand Climate Justice, 2024). This demand far exceeded the $300 billion per year by 2035 proposal later advanced by donor countries at COP29, which was sharply criticized for falling far short of the $1.46 trillion needed by 2030 and for favoring loans over grants (Climate Change News, 2024; UNCTAD, 2024).

Even the $100 billion per year pledge made in 2009 has repeatedly gone unmet. Worse, nearly 70% of what was disbursed in 2021 came as loans, not grants, deepening debt in countries already struggling to meet basic development needs (EcoAdvisors, 2020; IIED & International Indigenous Forum on Biodiversity, 2023; OECD, 2021; Oxfam International, 2023). These gaps are not merely financial; they are political. They undermine trust, reinforce asymmetries, and underscore the persistent underrepresentation of IPLCs in the governance and disbursement of climate and conservation finance (EcoAdvisors, 2020; Inclusive Conservation Initiative, 2024; OECD, 2021).

As subsequent chapters will show, effective land-sector mitigation depends not only on the rigor of global models or the volume of funding but also on the alignment of strategies with local governance systems, cultural priorities, and political-economic constraints. This chapter begins that analysis by examining how global narratives of climate ambition break down in practice.

Global South Demands for Climate Finance and the Credibility Crisis

As discussed earlier in this chapter, climate finance has become a flashpoint for recognized governance and justice in international mitigation. While donor governments often emphasize aggregate pledges, Global South leaders and civil society groups continue to point out that the scale, structure, and delivery of climate finance fall far short of what is needed or promised. These concerns are not new, but they have grown more urgent amid deepening crises and unmet expectations.

In 2024, the Conference of Non-Governmental Organizations in Consultative Relationship with the United Nations (CoNGO) again called for $5 trillion per year in climate-related finance to meet the combined costs of mitigation, adaptation, and loss and damage. This demand is echoed by movements such as the Global Campaign to Demand Climate Justice, which frames climate finance as both a legal obligation and a matter of historical accountability (Nacpil, 2024). Their position is grounded in economic analyses such as those by Hickel (2020), Chancel (2021), and Fanning and Hickel (2023), which quantify historical emissions and propose trillions in transfers as a condition for global equity.

The contrast with actual financial flows is stark. The $100 billion per year pledge made in 2009 has consistently gone unmet. As recently as 2021, nearly 70% of public climate finance was issued as loans, not grants, according to OECD data. And while donor countries proposed $300 billion per year by 2035 at COP29, this figure remains far below needs-based assessments and has been widely criticized as unambitious and politically convenient (Climate Change News, 2024; UNCTAD, 2024).

This financing gap has profound implications – not only for climate risk but also for whether land-based mitigation can be implemented effectively. Countries in the Global South often lack the technical, administrative, and institutional capacity to implement complex climate programs. This capacity gap is worsened by a political economy in which donor priorities often reflect risk aversion and geopolitical interests more than the governance realities or needs of recipients (Anozie, 2024; Sovacool et al., 2020).

The situation is particularly acute when it comes to IPLCs. As Dawson et al. (2021) observe, prevailing strategies frequently fail to engage IPLCs as true decision-makers. Without meaningful participation, institutional reform, and appreciation of the role of community lands (Climate Focus, 2022), even well-funded programs risk reinforcing exclusion and underperformance (Iocca & Fidélis, 2023).

What emerges is a broader crisis of credibility. Stakeholders across the Global South increasingly question whether international mitigation efforts, especially those driven by top-down financial logics and technical consultants, are capable of delivering real results. The disconnect between ambitious rhetoric, fragmented delivery, and limited local ownership is not just a bureaucratic failure. It reflects a fundamental misalignment between the way climate solutions are designed and the systems in which they must operate. Credibility cannot be engineered through financial pledges alone. It must be built through negotiation, inclusion, and grounded strategies that reflect IPLC leadership and context-specific realities.

Shortcomings of the Prevailing Paradigm

Despite a proliferation of land-sector mitigation initiatives – REDD+, NbS, Natural Climate Solutions (NCS), Agriculture, Forestry, and Other Land Use (AFOLU), and Payments for Ecosystem Services (PES) – the prevailing paradigm continues to prioritize carbon as the central metric (CBD, 2019; Harvey & Schonhardt, 2024; Nabuurs et al., 2022; Roe et al., 2021). As Busch (2015) observed, even well-intentioned REDD+ efforts face five persistent constraints, including structural incapacity and conflicting national priorities – challenges that remain largely unaddressed under current frameworks.

This carbon-centric logic reflects a deeper bias toward commodification and quantifiability – a logic that appeals to donors, consultants, and private investors but often misaligns with the sociopolitical and ecological realities of the places in which these programs are deployed (Beaury et al., 2024; Fletcher & Breitling, 2012; McAfee, 2015; Melanidis & Hagerman, 2022).

Nowhere is this bias more evident than in the reliance on standardized MRV systems. These frameworks prioritize measurable carbon outcomes while sidelining more meaningful indicators like secure tenure, local authority, and community capability (Haya et al., 2023; Sunderlin et al., 2014). Recent commentary from advocates of NbS attempts to clarify common misconceptions, emphasizing that when well-designed, NbS can support equity and biodiversity outcomes – yet many practical implementations fall short of this ideal (Nature4Climate, 2023; Seddon et al., 2020, 2021).

MRV systems create the illusion of scientific precision, but in practice, they depend on modeled counterfactual baselines – hypothetical scenarios of what would have happened without the intervention. These speculative baselines are often optimized to maximize credit generation rather than reflect plausible land-use trajectories (Dargie et al., 2017). As Karsenty et al. (2014) warn, this opens the door to rent-seeking and inflated carbon accounting (Havemann, 2009). Even when emissions reductions are recorded, deforestation may simply shift – or "leak" – to nearby areas, undermining the credibility of mitigation claims (Fearnside, 2017; West et al., 2020). Romm et al. (2025) argue, meanwhile, that carbon offsets are unfixable, and that more durable carbon dioxide removal (CDR) is what is needed.

The result is a system optimized not for productive land system change but for tradability and donor reassurance (Kim, 2023; Lang, 2023). MRV frameworks support investor and compliance reporting across dozens of jurisdictions but rarely foster the enabling conditions needed for long-term mitigation: secure tenure, trusted institutions, inclusive governance, and community ownership (Klooster,

2010). If emissions reductions occur, they are a byproduct of such conditions, not their precursor. Yet current programs invert this logic: They begin with global carbon targets and only later attempt to retrofit feasibility into contexts where tenure is contested, institutions are weak, or communities have been excluded from program design (Larson et al., 2013; Luo et al., 2024).

The consequences are not incidental. Many projects impose rigid land-use restrictions or top-down benefit-sharing schemes that are a poor fit with local governance systems or livelihood strategies (McDermott et al., 2013; Reed et al., 2017; Zagema, 2011). While modest benefits, like short-term employment or infrastructure, may be provided, they often serve as symbolic co-benefits rather than substantive empowerment (Agrawal & Redford, 2009; Brown, 2013; Büscher et al., 2022). The result is a pattern of disillusionment, passive resistance, or outright opposition.

Globally, REDD+ implementation continues to expose a deep gap between formal commitments to community engagement and the actual governance of mitigation projects, demonstrating how safeguards often fail to empower IPLCs in practice perpetuating disappointing results and outcomes for people and on a biophysical level (Cubas-Baez et al., 2025).

Over a decade ago, Brown's book *Redeeming REDD* warned that prevailing models failed to reflect the political and cultural conditions needed for success (Brown, 2013). Despite rebranding and rhetorical shifts under frameworks like NbS and NCS, those structural flaws persist. Appropriateness is the degree to which a strategy aligns with political, institutional, ecological, and cultural realities. While seemingly obvious, it remains poorly integrated into most climate mitigation and conservation design. Feasibility also suffers, particularly when programs rely on fragile institutions or superficial IPLC engagement (Garnett et al., 2018; Larson et al., 2022).

A more viable approach would invert the current logic. Rather than designing for carbon and attempting to "scale" through market mechanisms, land-sector mitigation should begin with what is socially and politically possible to achieve. This means addressing with tenure pluralism, overlapping claims, elite capture, administrative corruption, and institutional constraints, not ignoring them. Global recognition and technical finance should follow locally grounded strategies, not precede them. Veit and Reytar (2017) argue that land rights are not a co-benefit – they are the foundation for both climate and biodiversity outcomes. It is on that foundation that credible, just, and durable mitigation must be built.

The Illusion of Delivery: When Metrics Outpace Reality

A persistent and troubling gap exists between the ambitious goals set by global climate frameworks and the actual outcomes of land-based mitigation programs in the Global South (Boehm et al., 2023). This disconnect is not simply a matter of poor implementation. It is a structural flaw embedded in a paradigm that privileges technocratic modeling, commodified incentives, and financial metrics over the social, political, and institutional conditions that shape practical viability in complex institutional settings (Corbera et al., 2009; McAfee, 2015; Scheidel et al., 2020).

REDD+, NbS, NCS, AFOLU, and PES approaches all purport to offer "solutions." Unfortunately, the logic underpinning them remains rooted in theoretical constructs that frequently assume away the very problems they aim to solve. Key assumptions regarding baseline deforestation, leakage, additionality, and permanence are embedded in MRV protocols that are presented as neutral or objective (FAO, 2020). In practice, however, these protocols are shaped by methodological discretion, weak oversight, and market pressures that prioritize credit issuance over environmental or social integrity (Haya et al., 2023). What gets counted is what can be commodified, not necessarily what matters for long-term sustainability (Suckall et al., 2015). REDD+ in particular has failed to deliver promised transformation (Brockhaus et al., 2025; Lang, 2024).

Global mitigation strategies often present carbon reductions as universally transferable outputs that are measurable through standardized tools and tradable across contexts (Snyder, 2023). Yet translating these strategies into effective, inclusive action often breaks down in practice – particularly in ecosystem-based approaches – due to institutional fragmentation, unclear responsibilities, tenure insecurity, weak institutions, administrative corruption, and overlapping land claims (Davis et al., 2018; WWF, 2024), and lack of learning mechanisms (McDermott et al., 2019). These are not incidental challenges; they are structural features of the landscapes where mitigation is needed most. Instead of adapting to this complexity, many programs treat it as friction to be engineered away through top-down design, simplified metrics, and procedural shortcuts. Underlying this is a deeper, faulty, simplistic logic: that complexity itself is a flaw to be eliminated, rather than a condition to be understood and engaged. But land-sector mitigation without friction is not mitigation – it is avoidance dressed as progress.

This leads to what might be called the *illusion of delivery*. Governments sign pledges. Donors allocate funding. Developers design projects to meet emissions targets. But the underlying systems – local governance, legal recognition, IPLC authority – are seldom built to carry the burden. Implementation remains extractive, transactional, or symbolic, rather than participatory or transformative. And when challenges inevitably arise, the failure is attributed to local incapacity rather than poor structural design.

Most mitigation efforts are further compromised by a delivery logic dominated by intermediaries – technical consultants, international NGOs, or multilateral agencies – whose incentive structures prioritize compliance with donor protocols over alignment with local priorities as Gautam et al. (2021) present for Nepal, and Yazzie (2022) notes for NbS overall. In this world, reports, not relationships, judge projects by carbon credits, not trusted community leadership, respected local authorities, obtaining FPIC, or improved land use system metrics. A 2023 case in Papua New Guinea illustrates this dynamic: a coalition of environmental organizations publicly condemned the country's REDD+ regulation process as "deeply flawed," citing lack of meaningful participation, inadequate safeguards, and procedural opacity (Lang, 2023a). These issues exist within a broader landscape of structural challenges, including persistent tribal violence and insecure governance conditions (Muroa, 2022).

This gap between rhetoric and reality also manifests in policy language. Concepts like "inclusion," "participation," or even "co-benefits" have become ritualistically invoked to check normative boxes rather than to signal meaningful shifts in power (Savimbo, 2025; Schleicher et al., 2019). In many programs, FPIC is reduced to a procedural step rather than a substantive process of shared governance (Chapin, 2022; Daytec-Yañgot, 2021; United Nations, 2007).

As the rest of this chapter will explore, the prevailing model of land-sector mitigation continues to operate as if scalable, quantifiable, and transferable outcomes can be engineered independently of governance context. The evidence suggests otherwise. Until the model itself is recalibrated to begin with feasibility, contextual fit, and legitimacy, as judged by those on the ground, the gap between rhetorical ambition and implementation failure will persist.

As the Introduction emphasized, evaluating STAM frameworks – strategies, tools, approaches, and methods – requires more than technical rigor; it requires grounding in appropriateness, feasibility, and legitimacy as defined by those most directly affected. Almost invariably, these conditions are lacking in CMLSGS programming.

The Logic of Persistence: Why the Current Paradigm Remains Entrenched

Despite extensive critique and a lackluster performance record, the prevailing paradigm for CMLSGS has not only persisted but also expanded. Its endurance is not a testament to effectiveness but rather to the architecture of incentives that sustains it. Even mainstream economists acknowledge that climate policy is shaped more by political expediency than by optimal design (Helm et al., 2021).

This persistence is enabled by donor imperatives, bureaucratic preferences for standardized metrics, and the perceived elegance of carbon as a fungible, quantifiable unit. It is a system optimized not for social or ecological outcomes, but for risk-averse finance, accountability to funders, and technical auditability – criteria that international investors and institutions prioritize over long-term sustainability or governance reform (Ecosystem Marketplace, 2024; RRI, 2020).

Carbon, as a tradable commodity, fits neatly into the logic of global finance. It facilitates both voluntary and compliance-based offsets, enables streamlined MRV, and allows climate finance to move across borders without engaging politically sensitive issues like land tenure, distributive equity, or IPLC governance. In this way, carbon offers a veneer of action and measurability in a policy space riddled with social complexity and institutional ambiguity (Schroeder & McDermott. 2014; Turnhout et al., 2017).

But it is precisely this convenience that exposes the model's limitations. While mounting evidence shows that carbon market approaches often misprice risk, inflate benefit claims, and marginalize or exclude IPLCs, the global apparatus surrounding them continues to expand. This includes not only a proliferation of new methodologies and digital MRV platforms but also endorsements from high-profile initiatives such as the LEAF Coalition (Emergent, 2022), the IFC Performance Standards (IFC, 2012), and LandScale (2024), a multi-actor landscape sustainability assessment system developed by Verra, Rainforest Alliance, and partners.

Evidence reviews highlight that market-based approaches to mitigation and adaptation deliver mixed and often context-dependent outcomes (Beavor et al., 2024; Forest Peoples Programme, 2024; REDD-Monitor, 2023). Yet the system continues to reproduce itself. It is driven by actors whose financial, reputational, or institutional incentives align with the status quo rather than with transformational change.

The durability of the current model should not be mistaken for its legitimacy. It reflects a self-reinforcing regime, sustained by structural incentives that discourage engagement with the hard problems: land rights, IPLC authority, and governance transformation. Until these foundations are addressed, mitigation in the land sector will remain vulnerable to underperformance, resistance, and erosion of credibility.

Rebranding Without Reform: The Evasion of Transformational Change

What is systematically avoided in the current paradigm is not the rhetoric of transformation, but its implementation. True transformational change would require shifting authority to IPLCs, embedding negotiation and consent at the center of project design, and aligning mitigation objectives with local governance structures and development priorities. It would mean prioritizing appropriateness and feasibility, social as well as technical, at the landscape level, rather than treating them as afterthoughts.

Yet this shift is persistently deferred. Instead, familiar approaches are reintroduced under new branding – such as "jurisdictional REDD+," which, despite promises of scale and inclusivity, often replicates the limitations of earlier REDD+ models (Sylvera, 2023); "climate-smart landscapes," a term criticized for its vagueness and lack of clear community governance integration (Scherr et al., 2012); "nature-positive outcomes," flagged as susceptible to greenwashing and lacking accountability (University of Oxford, 2023); and "high-integrity carbon markets," whose standards remain inconsistently defined and prone to manipulation (Grist, 2024).

These rebrands preserve the core architecture of the prevailing model – centralized design, remote verification, and top-down finance – while presenting the illusion of reform. The same actors maintain control, and the same structural barriers go unchallenged.

This cycle reflects a kind of technocratic optimism detached from empirical evidence. Donors and implementing agencies regularly cite "lessons learned" in evaluations and conference statements, but those lessons rarely translate into structural reform. Governance models, benefit-sharing arrangements, and decision-making power remain largely unchanged. What results is a system that absorbs critique while resisting transformation (Robinson et al., 2018).

Rather than evolve, the system adapts by repackaging its tools and terminology to retain legitimacy in the face of mounting skepticism (Papillon et al., 2020). In doing so, it blunts the force of critique without confronting the political economy of exclusion, the need for IPLC leadership, or the structural redesign required for effective, durable climate mitigation. Because of this disconnect, momentum has grown around IPLCs' leadership, pointing to a possible turning point in global discourse – even as meaningful shifts in finance and governance remain slow (Pairojmahakij, 2023).

False Choices: Urgency, Inclusion, and the Transaction Cost Myth

One reason the current paradigm remains entrenched is the false binary often drawn between urgency and inclusion. Advocates for rapid climate action frequently claim that participatory processes are too slow, too complex, or introduce unacceptable risks of delay. Moreover, they argue that inclusion, particularly through FPIC and participatory MRV, raises transaction costs, which is viewed as a major obstacle to the efficiency and scalability of carbon markets (Cacho et al., 2013; Miller Thomson LLP, 2023).

But there is no evidence to support this claim. On the contrary, programs that lack legitimacy, grounded design, or community ownership are far more likely to fail through either active resistance or passive non-compliance. These are not ideological objections. They reflect lessons from decades of empirical research in conservation, development, and climate mitigation (Dawson et al., 2021, 2024; Garnett et al., 2018; Schuster et al., 2019).

The belief that inclusion slows down climate action flips the real problem on its head. It treats IPLCs as obstacles rather than essential partners and frames governance breakdowns as inefficiencies rather than design failures. Yet a growing body of evidence shows that IPLC-managed lands consistently outperform externally managed conservation areas, producing superior results in biodiversity protection, carbon sequestration, and long-term landscape resilience.

In short, the problem is not that participatory models are too slow. Even if such models introduce delays, the more important question is: What is the rationale for expediting processes to reduce transaction costs if the resulting interventions are destined to fail? Accelerating failure is not efficient; it is systemic blindness. The prevailing paradigm refuses to take inclusion seriously as the core enabling condition for success. As later chapters will argue, inclusion is not a trade-off against effectiveness. It is the foundation of any outcome that hopes to be just, durable, and scalable.

Toward a New Theory of Change: From Technocratic Pipelines to Negotiated Pathways

If the prevailing theory of change in land-based climate mitigation rests on technocratic delivery – design a project, certify its metrics, scale the model – then the alternative must begin by rethinking how change happens and who gets to define it.

For too long, the assumption has been that global carbon targets can be localized through performance incentives, safeguard frameworks, and external oversight. But this model treats social legitimacy as something that can be engineered rather than something that must be earned through negotiation. It presumes that the right indicators, measured by the right experts, will produce the right results – regardless of whether they align with the priorities, governance systems, or lived realities of the people whose land is in question.

What's needed instead is a theory of change built around negotiated governance: a process where IPLCs are not merely "included" but positioned as primary actors in shaping what mitigation means, what trade-offs are acceptable, and what success looks like in context (McElhinny et al., 2021; Tolo & Molnar, 2014). This does not reject science or climate urgency – it reframes them. Mitigation must be understood not as a pipeline to be optimized, but as a political and institutional process to be co-designed.

This also requires rejecting the logic of replication in favor of social alignment. What works in one jurisdiction may fail elsewhere not because it was poorly implemented, but because it was never legitimate or feasible to begin with. A credible approach to climate mitigation must allow for variability, learning, and compromise, grounded in the core enabling conditions discussed earlier.

IPLCs have not only demonstrated the capacity to manage ecosystems sustainably; they have often done so more effectively than external actors (Food and Agriculture Organization of the United Nations, 2022; Kothari, 2022; UNEP-WCMC, 2021). A global spatial analysis by Garnett et al. (2018) and more recent research by O'Bryan et al. (2021) found that Indigenous-managed lands overlap substantially with areas of high biodiversity and intact ecosystems and are as or more effective than formally protected areas at maintaining ecological integrity (IPBES, 2019; Tauli-Corpuz et al., 2020b). In Rwanda, for instance, studies have shown that tenure may be perceived as "secure enough" even without formal titles, reinforcing the opportunity for context-specific tenure solutions to the management of forest-rich commons (Brown & Hughes, 2017). This leads to the concept of Good Enough Land Tenure Security described in Chapters 4, 6, and 7.

Similarly, Dawson et al. (2021) synthesized a wide body of evidence showing that when rights are secure and governance structures respected, IPLC-led conservation delivers stronger ecological and social outcomes than top-down or externally imposed models. Evidence has long shown that community-based forest governance can deliver positive outcomes for both conservation and livelihoods (Agrawal & Chhatre, 2006).

Other studies reinforce this pattern: Schuster et al. (2019) found vertebrate biodiversity levels on Indigenous-managed lands in Australia, Brazil, and Canada equal to or greater than those in state-managed protected areas, while Sze et al. (2022) demonstrated that Indigenous lands play a key role in reducing deforestation and degradation across the tropics. Protected areas and conservation-driven mitigation initiatives have at times exacerbated poverty, particularly when local tenure and governance systems have not been adequately integrated into program design (Brockington & Wilkie, 2015).

What these findings suggest is not simply that IPLCs are capable actors in conservation, but that their governance systems, when properly supported, can outperform many conventional approaches to forest and biodiversity management. The failure lies not in IPLC capacity but in the persistent refusal of dominant mitigation frameworks to treat IPLCs as governing authorities rather than as stakeholders to be consulted (Conservation International, 2024).

This book proposes a new starting point: beginning with the people who govern the land, then building out policy, finance, and science in support of what they can agree is fair, feasible, and durable. As introduced in Chapter 1 and developed in detail in the Introduction, this theory of change reframes climate mitigation as a process grounded in legitimacy, co-developed authority, and locally defined priorities, not as a technocratic pipeline driven by external metrics and dubious assumptions.

Reframing Credibility: From MRV to Social License

The prevailing model of land-based climate mitigation in the Global South draws its credibility from methodological rigor, quantitative reporting, and third-party verification, embodied in standardized MRV systems. These frameworks are designed to quantify emissions reductions, typically against modeled counterfactual baselines, and lend an aura of scientific precision to what are ultimately speculative and discretionary judgments (West et al., 2024). While MRV systems have become the default standard for establishing credibility, they often obscure more than they reveal. Their focus on carbon data production frequently sidelines foundational conditions for sustainable land use: tenure security, governance capacity, and local legitimacy.

This bias toward MRV and carbon metrics is not merely technical but political. It privileges the authority of elite institutions, technical consultants, and international experts, who define the theory of change, metrics of success, and assumptions about effective mitigation. IPLCs are often invited to "participate" but rarely empowered to shape the architecture of programs or define success on their own terms. As a result, projects may generate impressive carbon data while eroding trust, disrupting livelihoods, or sparking land conflicts, especially when accountability systems are externally imposed without regard for local realities (Tauli-Corpuz et al., 2020a).

Empirical evidence further underscores the limits of this paradigm. Systematic reviews of nature-based interventions in the Global South find that most studies lack rigorous causal evidence and fail to link interventions clearly to climate mitigation or human well-being outcomes (Cheng et al., 2023; Romm et al., 2025). MRV frameworks frequently inflate carbon savings, downplay leakage, and sidestep the social and institutional conditions that determine long-term viability (Cheng et al., 2020; Haya et al., 2023). While the metrics may appear clear, their meaning is often hollow as they fail to credibly account for social development criteria.

This is not simply a technical flaw – it reflects a deeper structural distortion embedded in how credibility is defined and measured. MRV systems prioritize the production of carbon data over the realities that shape sustainable land use: tenure security, governance capacity, and local legitimacy. These foundational conditions are treated as background noise rather than as preconditions for success. As the Forest Peoples Programme (2024) and others have documented, stakeholder engagement is often superficial, benefit-sharing promises fall short, and development goals are routinely sacrificed to maintain carbon accounting integrity. IPLCs have far too often heard promises before – schools that never get built, jobs that don't materialize, and payments that are delayed or rerouted through opaque intermediaries.

Over time, these failures erode more than trust in individual projects – they chip away at the legitimacy of the entire model. Some communities respond through protest, litigation, or land repossession; others disengage more quietly, participating in form but withholding consent in substance. The result is a project that appears credible to donors, carbon registries, and international observers but fails in practice.

Even market insiders acknowledge the depth of these challenges. Calyx Global, a leading carbon credit rating agency, identifies persistent weaknesses in baseline methodologies, the difficulty of verifying additionality at jurisdictional scales, challenges in aligning nested project structures, and inconsistent benefit-sharing frameworks. These are not peripheral issues; they strike at the core of whether REDD+ and similar models can deliver either environmental integrity or social legitimacy (Calyx Global, 2024). These are not teething problems but structural flaws in a system built to serve external accounting needs rather than internal legitimacy. The claim that time will fix these issues often masks the fact that the same intermediaries benefit from deferring reform. In practice, the longer the system persists without fundamental change, the more deeply entrenched these distortions become. This has impacted REDD+ market share and value, as transaction volumes and prices have shrunk amid scrutiny and revisions to crediting methodologies (Ecosystem Marketplace, 2023).

At the root of this crisis is a set of unexamined assumptions: that reducing carbon emissions is the overriding priority – more urgent and fundable than tenure reform, rural development, or governance strengthening – and that global actors, not national governments or IPLCs, have the authority to define success. This produces a system of reverse subsidy, where the Global South provides land, political complexity, and institutional risk in exchange for carbon credits that are verified, monetized, and consumed elsewhere. What gets counted is carbon. What gets discounted is everything else.

A more grounded definition of credibility must center on social license: community approval based on negotiated terms, informed consent, and local governance authority. This goes beyond embedding "benefit sharing" into top-down projects. When IPLCs are involved from the start – defining the terms of land use, deciding whether carbon commodification is appropriate, and negotiating their own priorities – what emerges is not benefit sharing, but co-governance. The legitimacy of such arrangements stems not from the equitable distribution of pre-defined benefits, but from the consent-based construction of the initiative itself.

Without earned trust, what we refer to as social license, mitigation efforts are more likely to face resistance, stall out, or quietly fail. It is one of the key conditions, alongside legitimacy and contextual fit, that ultimately will make mitigation efforts scalable and sustainable.

References

Accountability Framework Initiative. (2020). Operational guidance on free, prior and informed consent: Requirements, best practices, and practical considerations for companies to fulfill their obligation to secure the FPIC of Indigenous Peoples and local communities (Minor revision: May 2020). Accountability Framework Initiative. Gland, Switzerland. https://accountability-framework.org/fileadmin/uploads/afi/Documents/Operational_Guidance/OG_FPIC-2020-5.pdf

Agrawal, A., & Chhatre, A. (2006). Explaining success on the commons: Community forest governance in the Indian Himalaya. *World Development*, 34(1), 149–166.

Agrawal, A., & Redford, K. (2009). Conservation and displacement: An overview. *Conservation and Society*, 7(1), 1–10.

Anozie, C. (2024). Left behind: Funding climate action in the Global South. *Texas A&M Law Review*, 11(2), 333–363. https://doi.org/10.37419/LR.V11.I2.2

Artelle, K. A., Zurba, M., Bhattacharyya, J., et al. (2019). Supporting resurgent Indigenous-led governance: A nascent mechanism for just and effective conservation. *Biological Conservation*, 240, 108284.

Baum, C. M., Fritz, L., Low, S., & Sovacool, B. K. (2024). Public perceptions and support of climate intervention technologies across the Global North and Global South. *Nature Communications*, 15(1), 46341. https://doi.org/10.1038/s41467-024-46341-5

Beaury, E. M., Smith, J., & Levine, J. M. (2024). Global suitability and spatial overlap of land-based climate mitigation strategies. *Global Change Biology*, 30(9), 2024–2039.

Beavor, A., Danaa, Z., Dowsett, L., Khan, S., Mokgano, E., & Prowse, M. (2024). *Evidence review on market-based approaches to mitigation and adaptation: Approach paper*. Independent Evaluation Unit, Green Climate Fund. https://ieu.greenclimate.fund/document/approach-paper-evidence-review-market-based-approaches-mitigation-and-adaptation

Bilal, A., & Känzig, D. (2024). *The Macroeconomic Impact of Climate Change: Global vs. Local Temperature*. National Bureau of Economic Research. Cambridge, MA 02138.

Boehm, S., Jeffery, L., Hecke, J., Schumer, C., Jaeger, J., Fyson, C., Levin, K., Nilsson, A., Naimoli, S., Daly, E., Thwaites, J., Lebling, K., Waite, R., Collis, J., Sims, M., Singh, N., Grier, E., Lamb, W., Castellanos, S., Lee, A., Geffray, M., Santo, R., Balehegn, M., Petroni, M., & Masterson, M. (2023). *State of Climate Action 2023*. Berlin and Cologne, Germany, San Francisco, CA, and Washington, DC: Bezos Earth Fund, Climate Action Tracker, Climate Analytics, ClimateWorks Foundation, NewClimate Institute, the United Nations Climate Change High-Level Champions, and World Resources Institute. https://doi.org/10.46830/wrirpt.23.00010

Brockhaus, M., Wong, G., & Moeliono, M. (2025, March 7). What have we learned from 15 years of REDD+ policy research? (Analysis). https://www.researchgate.net/publication/393081272_What_have_we_learned_from_15_years_of_REDD_policy_research_analysis

Brockington, D., & Wilkie, D. (2015). Protected areas and poverty. *Philosophical Transactions of the Royal Society B*, 370(1681), 20140271.

Brown, M., & Hughes, A. K. (2017). Is Land Tenure "Secure Enough" in Rural Rwanda? Paper presented at the Annual World Bank Conference on Land and Poverty.

Brown, M. I. (2013). *Redeeming REDD: Policies, Incentives and Social Feasibility for Avoided Deforestation*. Routledge.

Busch, J. (2015). Five hard truths for REDD+. *Environmental Science & Policy*, 55, 1–11.

Büscher, B., Koot, S., & Thakholi, L. (2022). Fossilized conservation, or the unsustainability of saving nature in South Africa. *Environment and Planning E: Nature and Space*, 5(3), 1455–1475.

Cacho, O. J., Lipper, L., & Moss, J. (2013). Transaction costs of carbon offset projects: A comparative study. *Ecological Economics*, 88, 232–243. https://www.sciencedirect.com/science/article/abs/pii/S0921800912004910

Calyx Global. (2024). A buyer's J-REDD+ quality checklist: Guidance for assessing jurisdictional REDD+ programs. *Calyx Global Research Hub*. https://calyxglobal.com/research-hub/research/a-buyers-j-redd-quality-checklist/

CBD. (2019). Post-2020 biodiversity framework: Nature-based solutions. *Convention on Biological Diversity*. https://www.cbd.int/doc/c/409e/19ae/369752b245f05e88f760aeb3/wg2020-05-l-02-en.pdf

Chancel, L. (2021). Climate change & the global inequality of carbon emissions, 1990–2020: Summary. *World Inequality Lab Working Paper 2021/21*. Paris School of Economics and Sciences Po.

Chapin, M. (2022, December 15). A challenge to conservationists: An interview with Mac Chapin. Current Conservation. https://www.currentconservation.org/a-challenge-to-conservationists-an-interview-with-mac-chapin-2/

Cheng, S. H., Costedoat, S., Sigouin, A., et al. (2023). Assessing evidence on the impacts of nature-based interventions for climate change mitigation: A systematic map of primary and secondary research from subtropical and tropical terrestrial regions. *Environmental Evidence*, 12, 21. https://doi.org/10.1186/s13750-023-00312-3

Cheng, S. H., McKinnon, M. C., Masuda, Y.J., Garside, R., Jones, K. W., Miller, D. C., et al. (2020) Strengthen causal models for better conservation outcomes for human well-being. *PLoS ONE*, 15(3), e0230495. https://doi.org/10.1371/journal.pone.0230495

Climate Action Network. (2024). US$5 trillion owed to Global South by Global North due to the climate crisis. *Climate Network*. https://climatenetwork.org/2024/09/20/us5trillion-owed-to-global-south-by-global-north-due-to-the-climate-crisis

Climate Central, World Weather Attribution, & Red Cross Red Crescent Climate Centre. (2024). Climate change and the escalation of global extreme heat: Assessing and addressing the risks. *A Look at Global Extreme Heat Over the Past 12 Months, How Climate Change Has Influenced This Heat, and Strategies to Prevent Increasingly Frequent and Intense Heat from Claiming Lives Worldwide*. https://www.climatecentral.org/report/climate-change-and-the-escalation-of-global-extreme-heat

Climate Change News. (2024). Fractious COP29 lands $300bn climate finance goal, dashing hopes of the poorest. https://www.climatechangenews.com/2024/11/23/fractious-cop29-lands-300bn-climate-finance-goal-dashing-hopes-of-the-poorest

Climate Focus. (2022, November). Sink or swim: How Indigenous and community lands can make or break national climate commitments. *Climate Focus and World Resources Institute report*. Retrieved from climatefocus.com.

Conservation International. (2024). Shifting the power: Empowering Indigenous Peoples and local communities for effective climate action. conservation.org.

Corbera, E., González, D., & Brown, K. (2009). Institutional dimensions of payments for ecosystem services. *Ecology and Society*, 14(1), 18.

Cubas-Báez, A., Sunderlin, W. D., Larson, A. M., & Frechette, A. (2025). From promises to practice: Persistent challenges in safeguarding local rights in forest carbon market initiatives. *CABI Reviews*, 2025, 0025. https://www.cabidigitallibrary.org/doi/10.1079/cabireviews.2025.0025

Dargie, G. C., et al. (2017). Age, extent and carbon storage of the central Congo Basin peatland complex. *Nature*, 542(7639), 86–90.

Davis, M. A., Rourke, M. L., Rohman, B., & Fargione, J. E. (2018). Should conservationists pay more attention to corruption? *Oryx*, 52(1), 1–5.

Dawson, N. M., Coolsaet, B., Bhardwaj, A., Brown, D., Lliso, B., Loos, J., Mannocci, L., Martin, A., Oliva, M., Pascual, U., Sherpa, P., & Worsdell, T. (2024). Reviewing the science on 50 years of conservation: Knowledge production biases and lessons for practice. *AMBIO*. https://doi.org/10.1007/s13280-024-02049-w

Dawson, N. M., Coolsaet, B., Sterling, E. J., Loveridge, R., Gross-Camp, N. D., Wongbusarakum, S., Sangha, K. K., Scherl, L. M., Phan, H. P., Zafra-Calvo, N., Lavey, W. G., Byakagaba, P., Idrobo, C. J., Chenet, A., Bennett, N. J., Mansourian, S., & Rosado-May, F. J.

(2021). The role of Indigenous Peoples and local communities in effective and equitable conservation. *Ecology and Society*, 26(3), 19. https://doi.org/10.5751/ES-12625-260319

Daytec-Yañgot, C. L. (2021). FPIC: A shield or threat to Indigenous Peoples' rights? In *Indigenous Peoples' Experiences on Free, Prior and Informed Consent: A Collection of Case Studies (Philippines)* (pp. 1–18). UN-REDD Programme. Accessed at: https://www.un-redd.org/sites/default/files/2021-09/Indigenous%20Peoples%20Experiences%20on%20Free,%20Prior%20and%20Informed%20Consent:%20A%20Collection%20of%20Case%20Studies(Philippines).pdf

Demand Climate Justice. (2024). Climate finance demands. *Global Campaign to Demand Climate Justice*. https://demandclimatejustice.org/tag/climate-finance

Dudine, A., & Szoke-Burke, S. (2020). Reimagining REDD+: A community-focused approach to REDD+ in Ghana. *Georgetown Environmental Law Review*, 33(3), 509–557.

EcoAdvisors. (2020). Conservation financing for conservation programs with Indigenous Peoples and local communities. *Prepared for the Nature Conservancy*. https://www.nature.org/content/dam/tnc/nature/en/documents/EA_IPLC_Full_Report_2020.pdf

Ecosystem Marketplace. (2023). *View from the Understory: State of Forest Carbon Finance 2023*. Washington, DC: Forest Trends.

Ecosystem Marketplace. (2024). Nature's investment frontier: Market developments in biodiversity and carbon credits. https://www.ecosystemmarketplace.com/publications/natures-investment-frontier/

Emergent. (2022). The LEAF coalition: Lowering emissions by accelerating forest finance. *Emergent Climate*. https://www.emergentclimate.com

Fanning, A. L., & Hickel, J. (2023). Compensation for atmospheric appropriation. *Nature Sustainability*, 6, 1077–1086. https://doi.org/10.1038/s41893-023-01130-8

FAO, 2020. Webinar: From reporting to payments: Current state and trends of REDD+ reporting to the UNFCCC. https://www.fao.org/redd/news/detail/zh/c/1316110/

Fawzy, S., Osman, A. I., Doran, J., & Rooney, D. W. (2020). Strategies for mitigation of climate change: A review. *Environmental Chemistry Letters*, 18, 2069–2094. https://doi.org/10.1007/s10311-020-01059-w

Fearnside, P.M. 2017. Deforestation of the Brazilian Amazon. In: H. Shugart (ed.) Oxford Research Encyclopedia of Environmental Science. Oxford University Press, New York, USA. doi:10.1093/acrefore/9780199389414.013.102

Fletcher, R., & Breitling, J. (2012). Market mechanism or subsidy in disguise? *Geoforum*, 43(3), 507–517.

Food and Agriculture Organization of the United Nations. (2022). *Forest Governance by Indigenous and Tribal Peoples*. FAO.

Forest Peoples Programme. (2024). REDD+ Alert: How the World Bank's jurisdictional REDD+ pilot in East Kalimantan discriminates against indigenous peoples. Retrieved from https://www.forestpeoples.org/publications-resources/reports/article/redd-alert-how-the-world-banks-jurisdictional-redd-pilot-in-east-kalimantan-discriminates-against-indigenous-peoples/

Garnett, S. T., Burgess, N. D., Fa, J. E., Fernández-Llamazares, Á., Molnár, Z., Robinson, C. J., . . . & Leiper, I. (2018). A spatial overview of the global importance of Indigenous lands for conservation. *Nature Sustainability*, 1(7), 369–374.

Gautam, D., Khatri, D. B., Lamichhane, B. R., & Poudel, B. S. (2021). The implication of REDD+ in community forestry (CF) governance and local livelihoods: A review from Nepal. *Contemporary Research: An Interdisciplinary Academic Journal*, 5(1), 24–34. https://doi.org/10.3126/craiaj.v5i1.40480

Grist. (2024). A federal attempt to foster 'high-integrity voluntary carbon markets' falls short, experts say New guidance for credit-based derivatives gives "imprimatur to a system that doesn't have credibility to begin with." *Grist*. https://grist.org/regulation/cftc-guidance-carbon-credit-offsets-derivatives/

Harvey, C., & Schonhardt, S. (2024, September 25). Carbon markets take center stage at climate week as criticisms abound. *E&E News*.

Havemann, T. (2009). Measuring the effectiveness of protected area management: A methodological review. *Sustainability*, 1(3), 583–594.

Haya, B. K., Alford-Jones, K., Anderegg, W. R. L., Beymer-Farris, B., Blanchard, L., Bomfim, B., Chin, D., Evans, S., Hogan, M., Holm, J. A., McAfee, K., So, I. S., West, T. A. P., & Withey, L. (2023). *Quality Assessment of REDD+ Carbon Credit Projects*. Berkeley Carbon Trading Project.

Helm, D., Hepburn, C., & Mash, R. (2021). *The Economics and Policy of Climate Change*. Oxford University Press.

Hickel, J. (2020). Quantifying national responsibility for climate breakdown: An equality-based attribution approach for carbon dioxide emissions in excess of the planetary boundary. *The Lancet Planetary Health*, 4(9), e399–e404.

Humpenöder, F., Popp, A., Schleussner, C. F., Orlov, A., Windisch, M. G., Menke, I., Pongratz, J., Havermann, F., Thiery, W., Luo, F., Jeetze, P. V., Dietrich, J. P., Lotze-Campen, H., Weindl, I., & Lejeune, Q. (2022). Overcoming global inequality is critical for land-based mitigation in line with the Paris agreement. *Nature Communications*, 13(1), 7453.

International Finance Corporation (IFC). (2012). *IFC Performance Standards on Environmental and Social Sustainability*. Washington, DC: International Finance Corporation. Available at: https://www.ifc.org/content/dam/ifc/doc/mgrt/ifc-performance-standards.pdf

IIED & International Indigenous Forum on Biodiversity. (2023). Indigenous Peoples and local communities excluded from decision making on money to help them tackle nature loss. *International Institute for Environment and Development*. https://www.iied.org/indigenous-peoples-local-communities-excluded-decision-making-money-help-them-tackle-nature-loss

Inclusive Conservation Initiative (ICI). (2024). Phase *Two Report: A Focus on Inclusive Finance*. Washington, DC: Inclusive Conservation Initiative. Published by Conservation International and the International Union for Conservation of Nature, with funding from the Global Environment Facility. Retrieved from https://www.thegef.org/newsroom/publications/inclusive-conservation-initiative-phase-2-report-focus-inclusive-finance

International Energy Agency. (2021). *World Energy Outlook 2021*. Paris: International Energy Agency. https://www.iea.org/reports/world-energy-outlook-2021

Intergovernmental Panel on Climate Change (IPCC). (2019). Climate Change and Land: an IPCC special report on climate change, desertification, land degradation, sustainable land management, food security, and greenhouse gas fluxes in terrestrial ecosystems [P.R. Shukla, J. Skea, E. Calvo Buendia, V. Masson-Delmotte, H.-O. Pörtner, D.C. Roberts, P. Zhai, R. Slade, S. Connors, R. van Diemen, M. Ferrat, E. Haughey, S. Luz, S. Neogi, M. Pathak, J. Petzold, J. Portugal Pereira, P. Vyas, E. Huntley, K. Kissick, M. Belkacemi, J. Malley, (eds.)]. *Intergovernmental Panel on Climate Change*. Available at https://www.ipcc.ch/srccl/

Intergovernmental Panel on Climate Change. (2021). Climate change 2021: The physical science basis. Contribution of Working Group I to the Sixth Assessment Report of the Intergovernmental Panel on Climate Change. Cambridge University Press.

Intergovernmental Panel on Climate Change. (2022a). Climate change 2022: Impacts, adaptation, and vulnerability. Contribution of Working Group II to the Sixth Assessment Report of the Intergovernmental Panel on Climate Change. Cambridge University Press.

Intergovernmental Panel on Climate Change. (2022b). Sixth Assessment Report, Working Group III – Summary for Policymakers.

Iocca, L., & Fidélis, T. (2023). Is there a place for Indigenous Peoples and local communities in climate change policy and governance? Learnings from a Brazilian case. *Land*, 12(9), 1647. https://doi.org/10.3390/land12091647

IPBES. (2019). *Global Assessment Report on Biodiversity and Ecosystem Services*. Intergovernmental Science-Policy Platform on Biodiversity and Ecosystem Services.

IUCN. (2019). *Nature-Based Solutions*. International Union for Conservation of Nature – Guidance document.

Jaiswal, S., Nagarajan, A., & Mythri, A. (2024). Projecting a food insecure world: Equity implications of land-based mitigation in IPCC mitigation pathways. *Environmental Science & Policy*, 155, 103724. https://doi.org/10.1016/j.envsci.2024.103724

Jang, N. (2024, August 7). For Nature-Based Solutions to Be Effective, We Need to Work with Indigenous Peoples and Local Communities. *International Institute for Sustainable Development (IISD) Insights*. https://www.iisd.org/articles/insight/nature-based-solutions-indigenous-peoples

Karsenty, A., Vogel, A., & Castell, F. (2014). "Carbon rights", REDD+ and payments for environmental services. *Environmental Science & Policy*, 35, 20–29. https://www.researchgate.net/publication/259522935_Carbon_rights_REDD_and_payments_for_environmental_services

Kim, T. (2023). Why market-based approaches to climate change won't save the planet. *Political Science Undergraduate Review*, 8(1). https://doi.org/10.29173/psur338

Klooster, D. (2010). Standardizing sustainable development? The Forest Stewardship Council's plantation policy review process as neoliberal environmental governance. *Geoforum*, 41(1), 117–129.

Kothari, A. (2022). The power of communities. *Environmental Governance*, 14(2), 145–152.

LandScale. (2024). LandScale: Powering Landscape Sustainability. *Rainforest Alliance, Verra, & Conservation International*. Retrieved August 19, 2025, from https://www.landscale.org.

Lang, C. (2023). What's wrong with carbon markets? *REDD-Monitor*. Retrieved April 18, 2025, from https://redd-monitor.org

Lang, C. (2023a). Papua New Guinea: Environmental alliance slams draft REDD+ regulation process as 'deeply flawed'. *REDD-Monitor Substack*. https://reddmonitor.substack.com/p/papua-new-guinea-environmentalalliance-90e

Lang, C. (2024, June 15). The Blattner family's Isangi REDD project in the Democratic Republic of Congo. *REDD-Monitor*. Retrieved from https://reddmonitor.substack.com/p/the-blattner-familys-isangi-redd

Larson, A. M., Brockhaus, M., Sunderlin, W. D., Duchelle, A. E., Babon, A., et al. (2013). Land tenure and REDD+: The good, the bad and the ugly. *Global Environmental Change*, 23(3), 678–689.

Larson, A. M., Dokken, T., & Atmadja, S. (2022). Forest tenure pathways to climate mitigation. *Proceedings of the National Academy of Sciences*, 119(46), e2200333119.

Lefstad, L., & Paavola, J. (2023). The evolution of climate justice claims in global climate change negotiations under the UNFCCC. *Critical Policy Studies*, 18(3), 363–388. https://doi.org/10.1080/19460171.2023.2235405

Lomborg, B. (2020). *False Alarm: How Climate Change Panic Costs Us Trillions, Hurts the Poor, and Fails to Fix the Planet*. New York: Basic Books.

Luo, C., Li, S., Hua, E., Forsell, N., & Chen, M. (2024). Synthesizing mitigation ambitions and implementation of the agriculture, forestry, and other land use (approp) sector. *Ecosystem Health and Sustainability*, 10, Article 0217. https://doi.org/10.34133/ehs.0217.

McAfee, K. (2015). The post- and future politics of green economy and REDD+. In B. Stephan, & R. Lane (Eds.), *The Politics of Carbon Markets* (pp. 237–260). New York and London: Routledge.

McDermott, C. L., Coad, L., Helfgott, A., & Schroeder, H. (2019). Operationalizing lessons from the ecosystem-based adaptation field. *Environmental Science & Policy*, 92, 87–95.

McDermott, C. L., Mahanty, S., & Schreckenberg, K. (2013). Examining equity: A multidimensional framework for assessing equity in payments for ecosystem services. *Environmental Science & Policy*, 33, 416–427.

McElhinny, V., Degawan, M., Dunne, P., & Cruz, A. (2021). *Indigenous Negotiations Resource Guide*. Arlington, VA: Conservation International.

Melanidis, M. S., & Hagerman, S. (2022). Competing narratives of nature-based solutions: Leveraging the power of nature or dangerous distraction? *Environmental Science & Policy*, 132, 273–284.

Miller Thomson LLP. (2023). The intersection of blockchain technology and carbon markets. https://www.millerthomson.com/en/insights/esg-and-carbon-finance/intersection-blockchain-technology-carbon-markets/

Muroa, R. (2022). Tribal violence in Papua New Guinea: Causes, impacts & interventions. *PNG NRI Discussion Paper 191.*

Nabuurs, G.-J., Mrabet, R., Abu Hatab, A., Bustamante, M., Clark, H., Havlík, P., House, J., Mbow, C., Ninan, K. N., Popp, A., Roe, S., Sohngen, B., & Towprayoon, S. (2022). Agriculture, forestry and other land uses (AFOLU). In IPCC, 2022: Climate Change 2022: Mitigation of Climate Change. Contribution of Working Group III to the Sixth Assessment Report of the Intergovernmental Panel on Climate Change.

Nacpil, L. (2024). Why the Global South is owed $5 trillion in climate reparations. *The Friday Times*. https://thefridaytimes.com/21-Sep-2024/why-the-global-south-is-owed-dollar-5-trillion-in-climate-reparations

Nature4Climate. (2023, September). Myth-busting: Addressing misconceptions about nature-based solutions for climate change. *Nature4Climate*. https://nature4climate.org/wp-content/uploads/2023/09/N4C-myth-busting-paper-FINAL-1.pdf

Newman, R., & Noy, I. (2023). The global costs of extreme weather that are attributable to climate change. *Nat Commun.*, 14, 6103. https://doi.org/10.1038/s41467-023-41888-1

O'Bryan, C. J., Garnett, S. T., Fa, J. E., et al. (2021). The importance of Indigenous Peoples' lands for the conservation of intact ecosystems. *Global Ecology and Conservation*, 24, e01405.

OECD. (2021). *Forward-Looking Scenarios of Climate Finance Provided and Mobilised by Developed Countries in 2021–2025: Technical Note*. Paris: OECD Publishing.

Oxfam International. (2023). Climate finance short-changed: The real value of the $100 billion commitment in 2019–20. *Oxfam Briefing Paper*.

Pairojmahakij, R. (2023, July 26). Are the tides turning for Indigenous Peoples and local communities? *RECOFTC*. https://www.recoftc.org/stories/are-tides-turning-indigenous-peoples-and-local-communities

Papillon, M., Rodon, T., & Lévesque, F. (2020). From consultation to consent: Oil sands development, Indigenous Peoples, and the limits of free, prior, and informed consent. *The Extractive Industries and Society*, 7(2), 341–349.

Reed, J., van Vianen, J., de Lima, M., Sunderland, T., & Ordway, E. M. (2017). Global outcomes of 15 years of Payments for Ecosystem Services. *Nature Sustainability*, 1(3), 143–150.

Rights and Resources Initiative. (2020). *Estimate of the Area of Land and Territories of Indigenous Peoples, Local Communities, and Afro-Descendants Where Their Rights Have Not Been Recognized: Technical Report. Washington*, DC: Rights and Resources Initiative. https://rightsandresources.org/wp-content/uploads/2020/09/Area-Study-Final-1.pdf

Robinson, B. E., Holland, M. B., & Naughton-Treves, L. (2018). Responding to climate change in forest management: Two decades of recommendations. *Forest Ecology and Management*, 419, 1–5.

Roe, S., Streck, C., Beach, R., Busch, J., Chapman, M., Daioglou, V., Deppermann, A., Doelman, J., Emmet-Booth, J., Engelmann, J., Fricko, O., Frischmann, C., Funk, J., Grassi, G., Griscom, B., Havlik, P., Hanssen, S., Humpenöder, F., Landholm, D., . . . Lawrence, D. (2021). Land-based measures to mitigate climate change: Potential and feasibility by country. *Global Change Biology*, 27(23), 6025–6058.

Romm, J., Lezak, S., & Alshamsi, A. (2025). Are Carbon Offsets Fixable? *Annual Review of Environment and Resources*, 50, 649–680. https://doi.org/10.1146/annurev-environ-112823-064813

Savimbo. (2025). Moving beyond a tokenistic participation of Indigenous Peoples in nature financing. https://www.savimbo.com/blog/moving-beyond-a-tokenistic-participation-of-indigenous-peoples-in-nature-financing

Scheidel, A., Temper, L., Demaria, F., & Martínez-Alier, J. (2020). Ecological distribution conflicts as obstacles to the UN sustainable development goals: Assessing the role of capitalism. *World Development*, 137, 105170.

Scherr, S. J., Shames, S., & Friedman, R. (2012). From climate-smart agriculture to climate-smart landscapes. *Agriculture & Food Security*, 1(1), 12. https://agricultureandfoodsecurity.biomedcentral.com/articles/10.1186/2048-7010-1-12

Schleicher, J., Zaehringer, J. G., Fastré, C., et al. (2019). Protecting half of the planet could directly affect over one billion people. *Nature Sustainability*, 2(12), 1094–1096.

Schroeder, H., & McDermott, C. (2014). Beyond carbon: enabling justice and equity in REDD+ across levels of governance. *Ecology and Society*, 19(1), 31. http://dx.doi.org/10.5751/ES-06537-190131

Schuster, R., Germain, R. R., Bennett, J. R., Reo, N. J., & Arcese, P. (2019). Vertebrate biodiversity on indigenous-managed lands in Australia, Brazil, and Canada equals that in protected areas. *Environmental Science & Policy*, 101, 1–6.

Seddon, N., Chausson, A., Berry, P., Girardin, C. A., & Turner, B. (2020). Understanding the value and limits of nature-based solutions to climate change and other global challenges. *Philosophical Transactions of the Royal Society B*, 375(1794), 20190120.

Seddon, N., et al. (2021). Getting the message right on nature-based solutions to climate change. *Global Change Biology*, 27(8), 1518–1546.

Snyder, K. A. (2023). Land-based mitigation strategies and their implications for local communities in the Global South. *Working Paper WP23KS1*. The Lincoln Institute of Land Policy.

Sovacool, B. K., Tan-Mullins, M., & Linnér, B. O. (2020). The political ecology of climate adaptation assistance: Why, where, and how international aid donors help satisfy basic needs. *Annals of the American Association of Geographers*, 110(2), 579–592.

Suckall, N., Tompkins, E. L., Stringer, L. C., & Vincent, K. (2015). Presenting triple-wins? Assessing projects that deliver adaptation, mitigation and development co-benefits in rural sub-Saharan Africa. *Ambio*, 44(1), 34–41.

Sunderlin, W. D., Larson, A. M., & Duchellę, A. E. (2014). Beyond the technical: The politics of developing the MRV system in Peru. *Carbon Management*, 5(3), 309–322.

Sylvera. (2023). An introduction to jurisdictional REDD+. *Sylvera*. https://www.sylvera.com/blog/an-introduction-to-jurisdictional-redd

Sze, J. S. S., Carrasco, L. R., Childs, D., & Edwards, D. P. (2022). Reduced deforestation and degradation in Indigenous lands pan-tropically. *Nature Sustainability*, 5(2), 123–130. https://doi.org/10.1038/s41893-021-00815-2

Tauli-Corpuz, V., Alcorn, J., Molnar, A., Healy, C., & Barrow, E. (2020a). Cornered by protected areas: Replacing 'fortress' conservation with rights-based approaches helps bring justice for Indigenous Peoples and local communities, reduces conflict, and enables cost-effective conservation and climate action. *World Development*, 130, 104923.

Tauli-Corpuz, V., et al. (2020b). *Local Biodiversity Outlooks 2: The Contributions of Indigenous Peoples and Local Communities to the Implementation of the Strategic Plan for Biodiversity 2011–2020 and to Renewing Nature and Cultures*. Moreton-in-Marsh, England: Forest Peoples Programme.

Tolo, V. A., & Molnar, A. (2014). Conceptualizing the role of Indigenous and local communities in nature conservation: A review of models and experiences. *Environmental Management*, 54(2), 167–183.

Turnhout, E., Waterton, C., Neves, K., & Buizer, I. M. (2013). Rethinking biodiversity: from goods and services to "living with". *Conservation Letters*, 6(3), 154–161. https://doi.org/10.1111/j.1755-263X.2012.00307.x

UNCTAD. (2024). Countries agree $300 billion by 2035 as new climate finance goal – what next? https://unctad.org/news/countries-agree-300-billion-2035-new-climate-finance-goal-what-next

UNEP. (2021). *Emissions Gap Report 2021*. United Nations Environment Programme.

UNEP. (2022). *Emissions Gap Report 2022: The Closing Window*. Nairobi: UNEP.

UNEP-WCMC. (2021). *State of Indigenous Peoples' and Local Communities' Lands and Territories: A Technical Review*. World Conservation Monitoring Centre.

UNFCCC. (2015). The Paris agreement. https://unfccc.int/process-and-meetings/the-paris-agreement/the-paris-agreement

United Nations. (2007). United Nations Declaration on the Rights of Indigenous Peoples. https://social.desa.un.org/sites/default/files/migrated/19/2018/11/UNDRIP_E_web.pdf

United Nations Development Programme. (2019). *Pathway for Increasing Nature-Based Solutions in NDCs: A Seven-Step Approach for Enhancing Nationally Determined Contributions through Nature-Based Solutions*. New York, USA: UNDP.

University of Oxford. (2023). *Greenwashing threat to achieving nature-positive world*. University of Oxford. https://www.ox.ac.uk/news/2023-09-22-greenwashing-threat-achieving-nature-positive-world

Veit, P., & Reytar, K. (2017). *Land Rights, Climate Change, and Biodiversity: Strategic Investments for Policy Impact*. World Resources Institute.

Wang, F., Harindintwali, J.-D., Wei, K., et al. (2023). Climate change: Strategies for mitigation and adaptation. *The Innovation Geoscience*, 1(1), 100015. https://www.the-innovation.org/article/doi/10.59717/j.xinn-geo.2023.100015

West, T. A. P., Bomfim, B., & Haya, B. K. (2024). Methodological issues with deforestation baselines compromise the integrity of carbon offsets from REDD+. *Global Environmental Change*, 87, 102863. https://www.sciencedirect.com/science/article/pii/S0959378024000670

West, T. A. P., Börner, J., Sills, E. O., & Kontoleon, A. (2020). Overstated carbon emission reductions from voluntary REDD+ projects in the Brazilian Amazon. *Proceedings of the National Academy of Sciences*, 117(39), 24188–24194. https://www.pnas.org/doi/10.1073/pnas.2004334117

WWF. (2024). *TNRC Final Report: Addressing Corruption's Impact on Conservation*. Washington, DC: World Wildlife Fund.

Yazzie, J. (2022). Indigenous people: Nature-based solutions won't save climate. *Atmos*. atmos.earth.

Zagema, B. (2011). Land and Power: The Growing Scandal Surrounding the New Wave of Investments in Land. Oxfam Briefing Paper 151, 51 pages. Oxford: Oxfam International. Retrieved from https://policy-practice.oxfam.org/resources/land-and-power-the-growing-scandal-surrounding-the-new-wave-of-investments-in-l-142858/

2 The Polarized Paradigm – Credibility, Control, and the Breakdown of Feasible Climate Action

This chapter builds on Chapter 1 by examining how political polarization, ideological divides, and entrenched institutional power further erode the credibility and feasibility of current mitigation strategies. As outlined earlier, effective land-sector mitigation depends on a set of core enabling conditions – appropriateness, feasibility, and legitimacy – which are routinely undermined by the technocratic and carbon-centric logic of the dominant paradigm. These are not just evaluative criteria; they are preconditions for the design and implementation of climate policies that aim to succeed in practice, not just on paper.

Land-sector mitigation efforts in the Global South are increasingly shaped not only by environmental urgency but also by deepening political polarization between North and South and within each. Disputes over historical responsibility, future obligations, and the legitimacy of mitigation instruments have fractured global climate negotiations. These divides reinforce a dominant paradigm that privileges technical control over social feasibility, obscuring lived realities and marginalizing those who steward much of the world's forests and biodiversity: IPLCs.

This polarization is neither accidental nor evenly distributed. On the one hand, the Global South continues to demand scaled-up mitigation finance, loss and damage compensation, and greater recognition of sovereignty in land-use decisions. On the other hand, the Global North remains fragmented. Within the United States and Europe, political movements prioritize national borders, economic anxiety, or cultural identity over collective climate ambition. In such contexts, climate change is often viewed as one issue among many, competing with rising debt, public safety, migration, and economic inequality (Carlisle et al., 2023; Cole et al., 2022). As a result, scientific urgency frequently clashes with political feasibility.

The result is a polarized and fragile consensus. As Roberts and Lamp (2021) argue in *The Six Faces of Globalization*, today's climate conflicts reflect divergent worldviews between global cooperation, national control, and justice-based claims. These clashing logics produce false binaries: decarbonization versus development, science versus sovereignty, and ambition versus delay. Caught in between are frontline communities whose rights and priorities are routinely sidelined.

This chapter examines how global polarization distorts climate governance and undermines the feasibility of mitigation efforts in the land sectors of the Global

DOI: 10.4324/9781003380320-3

South. It explores how shifting power dynamics, fractured science-policy narratives, and hierarchical governance models have created a legitimacy crisis.

Building on Chapter 1, it argues that unless credibility is redefined through local consent, inclusion, and appropriateness, the prevailing mitigation paradigm will continue to fail both ethically and operationally, as climate advocacy based on scientific urgency has too often been operationalized without social feasibility. This chapter traces how that disconnect is amplified by global political polarization, institutional weakness, and centralized control over land-based mitigation funding.

Global Political Shifts, Scientific Credibility, and the Enduring North–South Divide

The longstanding North–South divide in climate negotiations is rooted in decades-old tensions over equity, responsibility, and development priorities. Since the 1972 United Nations Conference on the Human Environment in Stockholm, developing nations have insisted that environmental protections must not undermine their right to develop (Najam et al., 2003). This tension has only intensified as the science of climate change has advanced and global mitigation targets have become more ambitious.

Recent years have seen this fragile consensus further destabilized by geopolitical instability and shifting domestic priorities. The Trump administration's 2020 withdrawal from the Paris Agreement and subsequent dismantling of global partnerships signaled a retreat from multilateral engagement, while the Biden administration's brief restoration of support – including a pledge to cut net greenhouse gas emissions by 61%–66% by 2035 (AP News, 2024) – was soon reversed by renewed isolationist policies (AP News, 2025). These policy swings have sent shockwaves through international climate finance, undermining donor trust and making recipients wary of investing political capital in externally driven mitigation regimes.

The broader outcome is a climate governance landscape marked by volatility and declining trust. Philanthropic strategies, private capital flows, and multilateral fund governance are all affected by these swings, leading to increased uncertainty and hesitancy among both donors and recipients.

At the same time, public trust in climate science itself has become deeply polarized. Apocalyptic rhetoric – often grounded in scientific findings – can backfire, fueling public fatigue, distrust, or ideological opposition (APA Climate Change Task Force, 2022; Büntgen, 2024; Fagan & Huang, 2019; Lomborg, 2020). In the United States, climate change is interpreted through partisan filters, with skepticism driven not by a lack of information but by "motivated reasoning" aligned with broader political identities (Druckman & McGrath, 2019; Hornsey et al., 2016; Hornsey & Fielding, 2020; Tesler, 2018; Van Boven et al., 2018). This results in a discourse where climate urgency is dismissed as alarmism and technocratic interventions are promoted as "neutral" solutions to deeply politicized problems (Allcott & Mullainathan, 2010; Lockwood, 2018).

This polarization is mirrored internationally: In the Global North, legislative gridlock and executive orders have undermined policy continuity, while in the

Global South, disillusionment with global institutions is widespread. The credibility crisis is thus not only technical but also political, stemming from contradictions in how science is framed, whose knowledge counts, and which actors are empowered to interpret risk. These contradictions are amplified when climate governance models ignore social realities and assume scientific authority can substitute for political legitimacy.

Despite incremental progress on direct funding and marginal reforms, the broader architecture of international climate governance remains fundamentally shaped by unresolved North–South tensions. At the center is the principle of common but differentiated responsibilities and respective capabilities (CBDR-RC), enshrined in the United Nations Framework Convention on Climate Change (UNFCCC) and reaffirmed in the Paris Agreement. CBDR-RC acknowledges the different historical contributions and present-day capacities of countries in confronting climate change, but it has not resolved long-standing disputes over who pays, who decides, and who benefits.

The Global South has consistently emphasized that the Global North, responsible for the vast majority of historical emissions, must provide climate finance that is predictable, transparent, and grant-based, rather than relying on loans or highly conditional mechanisms (Fanning & Hickel, 2023). Yet trust in these promises has steadily eroded. Actual disbursement lags far behind pledges, and key institutions are increasingly viewed as instruments of Northern influence rather than engines of climate justice.

Recent negotiations, such as COP29 in Baku, have underscored these disparities. Developed nations pledged $300 billion annually by 2035 to support adaptation and mitigation in developing countries – a figure far short of the $1.3 trillion estimated as necessary to meet climate goals. The G77 + China, representing more than 130 developing nations, called for at least $500 billion annually and proposed innovative mechanisms like a CO_2 tax and a global financial transaction tax to bridge the gap. The absence of several Global North leaders at COP29 only deepened concerns about the seriousness of their commitments (Le Monde, 2024).

Meanwhile, developing countries continue to face escalating climate impacts – droughts, GDP losses, migration, and food insecurity – while observing that global defense spending far outpaces investment in climate resilience (Adenle et al., 2017). These stark asymmetries have prompted growing calls to rethink global priorities and prevent what many describe as a looming humanitarian crisis.

The credibility of Global North leadership is further undermined by internal contradictions. Some countries tout net-zero targets while simultaneously investing in fossil fuel infrastructure or promoting carbon offset mechanisms that outsource emissions reductions to the South without transferring real power or resources. Such schemes risk replicating colonial logics under the banner of climate cooperation (Borras & Franco, 2018; Rights and Resources Initiative, 2023; Sovacool et al., 2022).

Ultimately, the Global North–South divide is about more than numbers. It is about narrative control, institutional design, and enduring power imbalances in global climate governance. As long as the market-dominated governance paradigm cloaks inequality in the language of science, neutrality, and market rationality, it will remain vulnerable to resistance and fail to generate the social license needed to drive meaningful change.

Conflict, Fragility, and the Diversity of Forested Landscapes in the Global South

While much attention is focused on dense tropical forests, a substantial share of climate action in the Global South occurs in drylands and other forested landscapes that meet FAO criteria for tree height and canopy cover. These are ecologically distinct from moist forests. Drylands cover about 40% of global land area, support two billion people, and contain over a quarter of the world's forests (FAO, 2021; Roe et al., 2021). These regions are highly vulnerable to climate change and land degradation. They also experience a disproportionate share of armed conflict and are characterized by state fragility, which undermines environmental governance and human security (Bristol Cabot Institute for the Environment 2024; Transparency International, 2021). Policies and programming in drylands in particular have habitually failed to appreciate how dryland communities have learned over centuries to productively harness the variability of their highly unpredictable environments (Gujo & Modiba, 2025; IIED, 2024).

Conflict dynamics in both dense forests and drylands are complex and highly context-dependent. In humid forests, armed groups may exploit timber and mineral revenues, use forests for cover, and drive displacement that increases pressure on forest resources (Harwell, 2010). In drylands, cycles of drought, resource scarcity, land degradation, and weak governance create perverse synergies, leading to conflict that accelerates environmental decline and forces migration while disrupting traditional governance systems (Bristol Cabot Institute for the Environment, 2024; FAO, 2021). Across both forest landscapes, climate change acts as a "threat multiplier," intensifying competition over land and water, eroding livelihoods, and increasing the risk of violence (IIED, 2024; Kyed & Chambers, 2023).

Effective climate action in these diverse forested landscapes requires approaches that are conflict-sensitive and also tailored to local ecological and socioeconomic realities. Integrating measures for climate adaptation, food security, and peacebuilding, especially through inclusive, locally led governance, can help address the interconnected challenges of conflict, fragility, and environmental change in both dense forests and drylands (FAO, 2021; World Bank, 2010). What this means is that mitigation around carbon sequestration or offset markets cannot be the principal driver of policies and programming.

Top-Down Governance Models

Many international organizations (IOs) design climate finance mechanisms, such as REDD+ and the Green Climate Fund (GCF), that prioritize national governments as intermediaries for distributing funds. While often justified as efficient for scaling up climate action, this model reinforces centralized control and systematically sidelines

IPLCs from meaningful participation in decision-making and benefit-sharing (Sovacool, 2021). Recent commentary has even argued that traditional aid models should be replaced by market-based incentives for forest conservation (Freund, 2025).

However, REDD+ initiatives, in particular, have faced significant criticism for casting IPLCs as passive recipients rather than active stewards or co-managers of forest resources (Global Forest Coalition, 2024). This is despite strong evidence that territories under Indigenous governance exhibit lower deforestation rates and better conservation outcomes (Bayrak & Marafa, 2016). Such exclusion is symptomatic of deeper structural biases in global mitigation systems that elevate technical expertise and carbon metrics over locally anchored legitimacy and traditional knowledge systems.

Structural barriers persist across multiple levels: Grassroots communities often lack direct access to mitigation funding, are excluded from formal policy spaces, and contend with inadequate FPIC procedures. Even well-meaning platforms, such as the IPs and LCs Voluntary Carbon Market Engagement Forum (Souza et al., 2023), remain limited in impact when embedded within frameworks that prioritize quantitative targets over participatory governance.

A persistent disconnect exists between scientific ambition and local political feasibility. While climate science plays a crucial role in articulating global risk, its authority is often misused to justify technocratic decisions that bypass local agency. Alarm-based modeling may accurately depict climate risks but is frequently deployed to support interventions that ignore complex land claims, fragile institutions, or contested governance landscapes. As Chapter 1 established, when scientific urgency is operationalized without local legitimacy, it becomes a tool of imposition rather than a guide for inclusion.

Current mitigation strategies, collectively comprising strategies, tools, approaches and methods (STAM), often fail to assess core determinants of feasibility such as tenure security, institutional strength, and socio-cultural dynamics. These oversights lead to implementation failures despite ostensibly rigorous modeling, and research shows that ineffective communication and mischaracterization of nature-based solutions can further undermine legitimacy and acceptance (Seddon et al., 2021). Credibility in such contexts cannot rest on technical precision alone (IUCN, 2020); it must be earned through political legitimacy and grounded in the lived realities of those most directly impacted (Kabisch et al., 2022; Lincoln Institute of Land Policy, 2024; Newell & Taylor, 2018; World Resources Institute, 2024).

Even frameworks that promote "science-based targets," as seen in REDD+ and NbS, often overlook the full breadth of relevant knowledge systems. These paradigms typically marginalize political ecology, anthropology, and Indigenous perspectives in favor of engineering-based metrics. This scientific reductionism fuels the broader social legitimacy crisis, producing interventions that may be technically sound but socially unsustainable (Seddon et al., 2020).

Market Mechanisms, Bureaucratic Barriers, and the Capture of Climate Finance

As outlined in Chapter 1, MRV frameworks often prioritize quantifiable carbon data over governance context, tenure clarity, and community credibility. While

Chapter 1 focused on the structural design flaws of MRV, this chapter explores how those systems are embedded within global finance mechanisms and donor-driven project pipelines that consolidate control away from IPLCs.

Market-oriented instruments – such as Payments for Ecosystem Services (PES), carbon trading, and REDD+ – have become increasingly favored by international institutions and donor agencies. While they may appeal to financiers and technocrats, they often conflict with the relational worldviews and governance traditions of local communities. These communities' stewardship of ecosystems is typically rooted in non-monetary values, kinship ties, and spiritual responsibility (Green-Win Project, 2018).

These forest carbon schemes are typically structured around intricate carbon verification systems that require legal status, technical acumen, and access to institutional infrastructure. Due to historic marginalization and chronic underinvestment, many customary landholders lack these prerequisites to take a lead. As a result, they are either sidelined entirely or compelled to participate via external intermediaries. This reproduces a system where communities execute externally defined agendas without meaningful influence over strategy, design, or evaluation (Schroeder, 2010).

The underlying assumptions of many emissions trading frameworks flatten ecological and cultural complexity. Modeling exercises often miscalculate carbon baselines, exaggerate co-benefits, or ignore plural land claims (Asiyanbi & Massarella, 2020; Brown & MacLellan, 2020; West et al., 2020). These approaches are also marked by epistemic bias – privileging economic logic over Indigenous governance norms and local, social legitimacy (Tauli-Corpuz et al., 2020). REDD+ in particular has been critiqued for its overreliance on technocratic assumptions about incentives and modeling, often in ways misaligned with the practical governance and ecological needs of frontline communities (Brown, 2013).

This problem extends into conservation knowledge systems more broadly. As Dawson et al. (2024) document, dominant scientific frameworks over the past five decades have repeatedly privileged top-down approaches to conservation and climate action, often marginalizing local perspectives and reducing the influence of Indigenous Peoples and Local Communities (IPLCs) in decision-making. When participatory processes – such as robust FPIC – are lacking or superficial, these dynamics can contribute to elite capture, community resistance, or conflict (AMAN, 2021; Scheidel et al., 2020), while also facilitating the transformation of ecological functions like carbon sequestration into tradable credits and investment products.

One of the most persistent barriers to effective climate action and equitable development in the Global South is the gap between policy rhetoric and reality on the ground. Nowhere is this more evident than in the Congo Basin, where corruption, opaque concession practices, and elite capture undermine both environmental integrity and the rights of IPLCs. Box 2.1 illustrates how these dynamics play out in the Cuvette Centrale peatlands with pressures for oil exploitation in one of the most sensitive global biomes for climate mitigation and carbon sequestration.

Box 2.1 Corruption, Policy Inconsistencies, and Possible IPLC Strategies in the Cuvette Centrale in the Democratic Republic of Congo (DRC)

The Cuvette Centrale peatlands of the Democratic Republic of Congo (DRC) have come to epitomize the disconnect between climate rhetoric and policy reality, drawing vocal opposition from civil society groups such as the Notre Terre Sans Pétrole campaign. Civil society campaign has come out vehemently against. While the DRC government positions itself as a "solution country" for global climate mitigation, it has simultaneously opened 52 new oil exploration blocks in the world's largest tropical peatland, threatening a critical carbon sink, local livelihoods, and biodiversity (Lang, 2025). This contradiction is emblematic of a broader pattern in the Congo Basin, where weak governance, endemic corruption, and opaque concession practices routinely undermine both climate and development goals (Africa Center for Strategic Studies, 2025; Global Witness, 2019).

The DRC and its neighbors consistently rank near the bottom of Transparency International's Corruption Perceptions Index, reflecting deeply entrenched systems of influence peddling, elite capture, and impunity for environmental crimes (Africa Center for Strategic Studies, 2025). Logging, mining, and oil concessions are often allocated through non-transparent processes, with politically connected actors benefiting at the expense of both the environment and IPLCs (CIFOR-ICRAF, 2025; Global Witness, 2019). This "extractive impunity" not only accelerates peatland degradation and carbon emissions but also erodes the rights, security, and food sovereignty of IPLCs who depend on these landscapes.

Is There a Responsible and Effective Strategy for IPLCs?

While democratic participation and advocacy at the ballot box are important, experience shows that local anti-corruption initiatives can also make a tangible difference. For example, the work of Innovative Resources Management (IRM) in Bandundu and Equateur provinces demonstrated that even in a context of endemic corruption, systematic and transparent local action can reduce *tracassaries* (petty corruption) along the Congo River. IRM's approach combined transparency, community education about the costs of corruption, creative problem-solving, and persistent compliance with agreed rules (Brown et al., 2004). Although IRM's work targeted river transport and "petty" corruption – of a different scale than the challenges posed by fossil fuel development – its principles of community vigilance, local coalition building, electoral process participation, and persistent engagement internationally and nationally offer food for thought for IPLCs facing renewed challenges posed by oil and logging in the Cuvette Centrale, as they showed that "it is possible to make significant reductions in corruption in a country as endemically corrupt as the DRC."

Ultimately, IPLCs will need to combine strategies:

- *Local action and monitoring* to expose and resist corruption and illegal concessions;
- *Coalition building* with civil society, media, and international allies to amplify their voices;
- *Demanding transparency, FPIC, and independent oversight* of all extractive projects;
- *Strategic use of legal and policy tools* to challenge illegitimate concessions and enforce rights.

While the structural obstacles are immense, the IRM experience shows that progress is possible, even in the DRC, when local communities are empowered, organized, and persistent.

Administrative hurdles further exacerbate exclusion. Applying for mitigation funding typically requires navigating grant portals, developing technical proposals, forecasting budgets, and submitting reports in donor languages – systems designed for formal organizations rather than community groups. As a workaround, many rely on large conservation NGOs like WWF, WCS, Conservation International, and IUCN. These BINGOs act as financial conduits, controlling vast funding portfolios. As of 2024, WWF managed over $135 million in GCF projects, while IUCN oversaw more than $330 million in ecosystem-based grants. Still, direct financial flows to grassroots actors remain under 10% (Buchner et al., 2021; Climate Policy Initiative, 2021; IIED, 2020).

While these intermediaries bring technical capacity, they also entrench bureaucratic dominance. They often shape project pipelines, determine evaluation metrics, and filter reporting for international audiences. Communities are thus relegated to implementation roles, unable to shape the agenda or redefine success. This can result in technically successful offset programs that nevertheless erode trust, ignore tenure pluralism, or fail to produce lasting outcomes. Oversight frameworks routinely inflate carbon gains while overlooking the institutional fragilities that ultimately determine durability (Cheng et al., 2023).

The central flaw lies in a belief that data can substitute for consent and that metrics can stand in for earned legitimacy at community and higher jurisdictional levels. True accountability emerges not from consultant checklists but from governance structures rooted in social trust, local oversight, and negotiated terms.

Fortunately, alternatives are emerging. Simplified, multilingual application platforms – like those under the GCF's Enhanced Direct Access model – have begun reducing procedural barriers. Flexible, community-led grant models, such as the Dedicated Grant Mechanism (DGM), demonstrate how direct mitigation funding aligned with local priorities can reduce deforestation more effectively than external interventions (World Bank, 2020). Territorial accountability boards in regions like Mesoamerica offer locally governed mechanisms for evaluating proposals based on FPIC compliance and cultural fit.

These are not just better administrative tools – they represent a rebalancing of power. When traditional stewards gain control over financial instruments, they gain influence over the systems that shape land use. Without this shift, mitigation efforts will continue to prioritize donor optics over grounded outcomes, and the credibility gap between model and reality will only grow wider.

Power and Agenda-Setting by Conservation Intermediaries

BINGOs such as The Nature Conservancy, Conservation International, WWF, WCS, Re:wild, Pew Charitable Trusts, and IUCN, exert outsized influence over both the climate and conservation policy landscape. Their privileged access to major donors, seats at global negotiation tables, and control over project design pipelines and "best practice" in CMLSGS, allow them to shape which initiatives are considered "bankable," "high integrity," or scalable. They also set technical frameworks that define how success is measured and who is qualified to lead.

A prominent example is the 2024 alliance among six major BINGOs to develop sovereign debt-for-nature swaps – a mechanism projected to channel $100 billion into conservation finance (Conservation International, 2024). Though marketed as innovative and transformative, these deals reinforce the central authority of large NGOs to determine priorities, manage resources, and define the terms of engagement. Without binding requirements for IPLCs to co-design or approve such efforts, these initiatives risk perpetuating exclusion under the guise of climate innovation.

Through platforms like results-based finance, biodiversity offsetting, debt-for-nature swaps, and "nature-positive" investment frameworks, BINGOs help institutionalize the technocratic climate governance paradigm – what this book terms the CMLSGS model. This often sidelines community governance systems and reinforces elite control over conservation priorities. Büscher and Fletcher (2020) argue that these arrangements elevate market metrics and managerial expertise over endorsement through negotiated governance, consent, and local knowledge.

The consequences are visible on the ground. BINGO-led projects have, in multiple cases, been linked to land dispossession, elite capture, and human rights abuses – particularly where FPIC is weak or ignored (Rights and Resources Initiative, 2018; Scheidel et al., 2020). In these contexts, even well-meaning interventions may replicate extractive patterns under a green veneer.

Correcting these power asymmetries requires more than improving access or adding consultation checkboxes. What's needed is a governance realignment: one that embeds IPLC-defined priorities and knowledge systems into the institutional frameworks that determine funding flows and project design and gives IPLCs real decision-making authority – including veto power – over conservation investments that affect their territories (Garnett et al., 2018; Pacific Institute, 2022; Tengö et al., 2017).

As discussed further in Chapter 5, the current finance architecture remains optimized for risk-averse donors and public relations outcomes, often reinforcing mitigation public relations campaigning that is more concerned with integrity and virtue signaling than delivering it. But consent-based governance frameworks cannot be simulated. It must be built through institutional design that redistributes power and affirms Indigenous governance as foundational to effective land-sector climate action. Addressing this power asymmetry requires not only reforming how climate finance is accessed but also rethinking who sets agendas, defines impact, and exercises oversight.

Direct Funding Mechanisms: Pathways to Equity and Agency

While systemic reform remains essential to address the structural inequities in climate finance, interim measures can enhance equity and effectiveness within the current architecture. Direct funding mechanisms – which bypass intermediaries like national governments and BINGOs – offer a critical pathway to reduce bureaucratic barriers and center IPLCs' agency in climate action. These mechanisms are not just procedural tweaks; they represent a foundational shift toward redistributing power and legitimizing IPLC leadership.

Simplified Access

Complex application processes routinely exclude IPLCs from accessing climate funds. Multilingual portals and streamlined procedures, such as those piloted by the GCF's Enhanced Direct Access modality, allow communities to apply for funding without relying on intermediaries. For example, the Indigenous Peoples' Funding Window under the UN Development Programme reduces administrative hurdles by accepting proposals in local languages and prioritizing community-led diagnostics. Such reforms affirm self-determination by treating IPLCs as partners, not petitioners, in climate governance.

Flexible, Long-Term Grants

Short-term, project-based funding often forces IPLCs to conform to donor priorities rather than their own. Flexible grants tied to community-defined agendas, like those administered through the DGM, have demonstrated success in bridging this gap. In Brazil and Indonesia, DGM grants enabled IPLCs to design 5 to 10-year forest management plans aligned with traditional knowledge systems, resulting in a 30% reduction in deforestation compared to top-down projects (World Bank, 2015). Scaling such models requires funders to shift from rigid logframes to adaptive financing that trusts IPLCs to allocate resources based on evolving needs.

IPLC-Led Accountability

Traditional monitoring systems prioritize carbon metrics over social legitimacy. Community-governed review boards – such as those implemented in Mesoamerica's Territorial Funds – flip this dynamic. These boards assess proposals for FPIC compliance, cultural appropriateness, and long-term sustainability. In Guatemala, a Maya-led board rejected a REDD+ project that lacked community consultation, redirecting funds to a community-managed agroforestry initiative that reduced emissions while strengthening food security (RRI, 2022).

Case in Point: The Dedicated Grant Mechanism

The DGM, part of the World Bank's Forest Investment Program, exemplifies how direct funding can transform outcomes. By channeling $80 million directly to IPLC organizations across 12 countries, the DGM supported initiatives ranging from forest patrols in the Congo Basin to seed banks in the Peruvian Amazon. Independent evaluations found that DGM projects achieved higher cost-effectiveness and lower conflict rates than comparable BINGO-led programs, largely because communities controlled budgeting and implementation (World Bank, 2020).

Beyond Finance: Redistributing Power versus Overthrowing Capitalism

These mechanisms are not merely about moving money; they are about redistributing negotiating power. When IPLCs manage climate finance directly, they are better positioned to act as credible institutional partners, not symbolic stakeholders, but actors with grounded leverage. In Ghana, for example, communities using direct access grants established a precedent requiring private investors to seek their consent before acquiring land for carbon projects (IUCN, 2023). While such steps do not dismantle the prevailing hierarchies that govern global climate finance, they create "proto-platforms" for more accountable and negotiated governance, as discussed in Chapter 7.

At the other end of the political spectrum, some voices within the Global North climate movement – most notably led by journalist and activist George Monbiot – have argued that the only meaningful response to ecological collapse is to "overthrow capitalism" (Monbiot, 2019). This view reflects a growing disillusionment with the structural failures of market-led systems and the ecological harm they enable. In fact, according to the 2020 Edelman Trust Barometer, 56% of respondents across 28 countries agreed that "capitalism as it exists today does more harm than good in the world" (Edelman, 2020). Gallup polling has also shown that among US adults aged 18–29, support for socialism has at times outpaced support for capitalism (Gallup, 2021).

Yet despite these shifts in perception, most people are not calling for a full-scale systemic rupture. The global trend is not one of revolutionary rejection, but rather one of dissatisfaction with current outcomes – paired with an appetite for reform, redistribution, and structural accountability. Monbiot's critique may reflect a legitimate moral frustration, but it remains politically improbable as a strategy for mobilization on a global scale.

Meanwhile, many BINGOs – also primarily based in the Global North – have moved in the opposite direction, aligning themselves with private sector actors and market-driven frameworks like carbon trading. These alliances have often marginalized IPLCs and, arguably, contributed to underwhelming outcomes for both biodiversity and climate mitigation. What emerges is a fractured climate politics within the Global North itself: on one side, revolutionary anti-capitalist rhetoric; on the other, technocratic NGO–private sector partnerships, neither of which adequately centers the rights and leadership of the communities who actually steward the land.

A different path is needed. One that rejects both technocratic reformism and calls for sweeping, systemic transformation often found in critical climate (Klein, 2014; Schulz & Siriwardane, 2015). It calls for reduced paternalism, the institutionalization of negotiated governance, and IPLC leadership rooted in social and political feasibility. In doing so, it aims to move beyond symbolic participation toward structures that generate durable legitimacy and shared authority.

Toward a Transformative Model: IPLC-Led Negotiation Platforms

A truly sustainable approach to climate mitigation in the land sector requires re-centering power through IPLC-led negotiation platforms. As outlined in Chapter 7, this model repositions IPLCs – not donors, governments, or intermediaries – as the principal architects of land-based mitigation strategies. It begins by securing land tenure and building analytical and negotiation capacity, enabling communities to enter into agreements as equals. At the core of this model are Jurisdictional Negotiation Platforms (JNPs) – institutions designed for structured engagement between IPLCs, governments, and private sector actors.

These platforms are not merely consultative spaces. They are institutional innovations that enable IPLCs to define priorities, assess trade-offs, and manage resources on their own terms. Financing will flow directly into these processes, not around them – anchored in FPIC, transparency, and local legitimacy.

This approach also calls for simplification of application processes and targeted technical support to ensure that IPLC organizations can access funding independently. Rather than treating governance as a technical challenge to be solved through outside expertise, the paradigm shift positions IPLCs as governance partners whose consent, knowledge, and vision define the terms of engagement through providing social license and the specifics of negotiated agreements themselves (Lemos et al., 2021).

While incremental reforms help reduce harm, this model advances a deeper realignment – one that treats governance transformation not as a byproduct of mitigation, but as its precondition.

Inclusive Governance Structures

Governance challenges in the Global South are often rooted in institutional weaknesses – such as corruption, lack of transparency (Transparency International, 2021), and exclusionary decision-making. As Acemoglu and Robinson (2012) argue in *Why Nations Fail*, extractive institutions limit inclusive participation and inhibit equitable development. These same dynamics undermine climate resilience and the feasibility of land-based mitigation.

Adaptive governance frameworks – characterized by multi-level coordination, iterative learning, and inclusive participation – have been proposed as a solution (Chaffin et al., 2014; Sierra & Suárez-Collado, 2021). Yet in practice, these frameworks often remain advisory or symbolic.

For CMLSGS to succeed, governance must embed IPLC agency directly into institutional design. As detailed in Chapters 6 and 7, this requires IPLC-led negotiation platforms that link communities to governments and private entities. These platforms give IPLCs an actual seat at the table, allowing them to shape policy, allocate resources, and define success on their own terms.

Parallel reform is also needed within IOs. For example, the UNFCCC could establish a permanent Indigenous Peoples' Advisory Board with formal review and veto powers over policies that impact IPLC territories. Enforcement of FPIC across all climate finance mechanisms must be strengthened to close the credibility gap.

Without structural reforms, inclusive rhetoric will continue to mask extractive practices.

Other Effective Area-Based Conservation Measures: Opportunities and Caveats

In contexts where climate and conservation programming are just getting underway – and before robust governance structures or tripartite negotiation platforms are in place – Other Effective Area-Based Conservation Measures (OECMs) have emerged as a potentially useful category. OECMs are designed to recognize and support territories that deliver long-term conservation outcomes outside formal protected areas. Under the right conditions, they can serve as flexible, interim governance tools that support IPLC stewardship, territorial integrity, and adaptive management without relying on centralized exclusionary practices (CBD, 2018).

Yet theory often diverges from practice.

Critics have warned that without rigorous safeguards and IPLC governance, OECMs risk becoming a "catch-all" designation for state or NGO land control efforts. These efforts can sidestep FPIC and misattribute IPLC-led conservation as externally designed success stories. As the Forest Peoples Programme (2021) warns, this risk is particularly acute when OECMs are applied hastily to meet external targets like the global 30×30 pledge, which aims to protect 30% of the world's land and sea by 2030.

The Convention on Biological Diversity not only defines OECMs as areas "governed and managed in ways that achieve positive and sustained long-term outcomes for the in situ conservation of biodiversity" but also emphasizes that recognition of these areas must reflect the rights, knowledge systems, and governance priorities of IPLC (CBD, 2018). Despite this guidance, OECM designations have been used by some governments and conservation organizations in ways that sideline IPLC voices or reinforce existing hierarchies. A recent report by the Rights and Resources Initiative (2023) underscores that expanding OECMs without simultaneously addressing land rights and tenure security may reproduce the same land appropriation dynamics that plagued older protected area models.

This raises a set of familiar but unresolved questions. Who defines conservation success? Who monitors compliance, and who benefits from recognition? And more fundamentally, who holds the power?

In fragile or transitional governance settings, OECMs may offer a provisional path forward. Used transparently and with full IPLC participation, they can create breathing room for communities to engage in adaptive conservation while broader governance platforms are being developed. But used as shortcuts – driven by deadlines or donor pressure – they risk becoming another technocratic repackaging of enclosure.

Like many tools in the prevailing mitigation paradigm, the label itself means little. The critical test is whether the underlying process is anchored in Indigenous- and local community-defined priorities, meeting appropriateness and feasibility criteria.

Reframing Credibility: From MRV to Social License

The prevailing model of land-based climate mitigation derives credibility from methodological rigor, quantitative reporting, and third-party verification – embodied in standardized MRV systems. Designed to quantify emissions reductions against modeled counterfactual baselines, these frameworks lend scientific precision to speculative judgments (Karsenty, 2019). While MRV systems dominate credibility assessments, they often obscure more than they reveal, sidelining foundational conditions for sustainable land use: tenure security, governance capacity, and local legitimacy.

This bias is not merely technical but political. MRV privileges the authority of elite institutions and technical consultants who define success metrics, marginalizing IPLCs. Projects may generate impressive carbon data while eroding trust, disrupting livelihoods, or sparking conflicts, particularly when accountability systems are imposed without regard for local governance. As noted in Chapter 1, what is missing is not more data, but alignment with the core enabling conditions introduced earlier.

REDD+ in Ghana provides a clear illustration. Implementation of REDD+ has been hindered by limited funding, fragmented donor support, and a complex, pluralistic tree tenure system that fails to incentivize farmers to protect or plant trees.

Under Ghana's constitution, all naturally occurring trees are vested in the government, meaning that farmers do not benefit directly from maintaining shade trees on their land, an arrangement that undermines local stewardship and has limited the effectiveness of project-based REDD+ approaches. As a result, Ghana has shifted toward jurisdictional REDD+ models, which allocate funding and incentives at the ecological region or national level. This approach aims to align incentives, improve benefit-sharing, and address governance challenges such as elite capture and unclear carbon rights, but progress remains slow due to ongoing tenure and accountability issues (Climate and Land Use Alliance, 2015; CSIR-FORIG, 2020; Dudine & Szoke-Burke, 2020).

Empirical evidence underscores the limits of this paradigm. Systematic reviews reveal that most nature-based interventions lack causal evidence linking them to climate or well-being outcomes (Beaury et al., 2024; Chausson et al., 2020; Cheng et al., 2023). MRV frameworks frequently inflate carbon savings, downplay leakage, and ignore social-institutional conditions for viability (Haya et al., 2023).

A similar pattern is evident in Papua New Guinea: The REDD+ Readiness Phase 2 project (2018–2020) focused on building the capacity of government agencies and stakeholders, developing a comprehensive national REDD+ strategy, and enhancing the country's forest monitoring and safeguards systems. This phase included the establishment of a National REDD+ Steering Committee, technical working groups, and a national communications strategy to ensure broad stakeholder engagement, including provincial governments, civil society, and Indigenous landowners. The project's achievements, such as the development of a Five-Year National REDD+ Finance and Investment Plan and the implementation of safeguards information systems, have laid the groundwork for PNG to access results-based payments under the UNFCCC; however, unresolved issues around land tenure, benefit-sharing, and inclusive participation remain key challenges for scaling up REDD+ implementation (Legrand & Gowae, 2020; UNDP Papua New Guinea, 2021).

MRV frameworks' structural flaws are also evident in Mesoamerica. Community forest concessions in the Maya Forest – particularly in Guatemala's Petén region – have demonstrated significant success in reducing deforestation rates compared to non-concession areas. Local communities, empowered with legal rights to manage forest resources, have played a central role in monitoring, law enforcement, and sustainable timber harvesting, resulting in lower rates of illegal logging and greater biodiversity conservation.

The Maya Forest case underscores that when IPLCs have secure tenure and genuine authority, they can deliver both climate and development benefits more effectively than top-down or externally managed programs (Radachowsky et al., 2012; RRI, 2022). Yet such success stories remain the exception rather than the rule.

In most contexts, repeated failures – stemming from exclusion, poor governance alignment, and technocratic overreach – have chipped away at trust not just in individual projects but also in the broader paradigm of land-based climate mitigation. Familiar approaches are then reintroduced under new branding – "jurisdictional REDD+" (Proforest Initiative, 2023), "climate-smart landscapes" (Bangalore et al., 2016), "nature-positive outcomes" (World Wide Fund for Nature, 2021), and "high-integrity carbon markets" (ICVCM, 2023) – often without substantive changes to governance, financing flows, or community authority. Rebranded solutions often fail because they do not meet the foundational design preconditions for local legitimacy and feasibility.

The problem, however, is structural, not technical. MRV systems prioritize carbon data over the realities of tenure pluralism, elite capture, and institutional fragility. A self-reinforcing network of carbon standards, validators, and consultancies perpetuates this model, prioritizing tradability and donor confidence over legitimacy (Ecosystem Marketplace, 2023).

As discussed in Chapter 1, MRV systems prioritize abstraction over grounded realities. What matters here is how those systems are institutionally protected – enshrined by donor protocols, consultancy firms, and registries that rarely engage with the governance risks they obscure.

A credible alternative must center on social license – community approval grounded in negotiated terms, informed consent, and co-governance. This contrasts sharply with "benefit sharing," which presupposes external control. When IPLCs define land-use terms and priorities from the start, legitimacy stems from consent-based design, not pre-defined benefit distribution. In Rwanda, "good enough" context-specific tenure solutions – even without formal titles – have enabled sustainable forest management where rigid MRV frameworks failed (Brown & Hughes, 2017).

Without social license, climate mitigation efforts often falter, are quietly resisted, slowly unravel, or are simply ignored. Credibility doesn't come from metrics alone. It has to be earned through trust, negotiation, and governance that centers IPLC rights, knowledge, and consent.

What this chapter has shown is that the crisis facing land-based mitigation in the Global South is not just about flawed technical models or underperforming projects. It's rooted in deeper problems: political polarization, broken promises in international finance, and a dominant approach – what this book calls the CMLSGS paradigm – that puts carbon accounting and institutional control ahead of appropriateness, feasibility, and social legitimacy.

This paradigm has produced systems and processes where IPLCs are praised in theory but sidelined in practice. And while IPLCs are not a magic solution, the evidence is clear: When they have real rights and real power, climate and conservation outcomes improve. Including their leadership isn't just fair, it's necessary.

Many current strategies, including REDD+, carbon markets, and conservation finance for NbS, are driven by the need to demonstrate externally defined metrics managed by intermediaries with vested marketing interests. But their effectiveness is often superficial, and their credibility wears thin when they ignore the political and cultural realities on the ground. Goals like "30×30" will fail unless IPLC consent and leadership are treated as fundamental, not optional.

The way forward isn't to simplify the problem but to meet its complexity with better tools. That means shifting away from centralized climate finance models, market-driven governance, and rigid performance frameworks and toward locally grounded platforms where communities help shape the rules. IPLC-led negotiation spaces, direct access to finance, and genuine accountability are not just policy tweaks – they are the foundation of a more legitimate system.

If mitigation efforts are to succeed, they must be reshaped to align with the social, political, and institutional realities they intend to serve. Without grounding strategies in the core enabling conditions of local legitimacy, feasibility, and contextual fit, the credibility gap will only widen, regardless of how urgently NbS, NCS, and other favored approaches are marketed.

Chapter 3 continues this critique by examining how the dominant suite of land-based mitigation strategies – including REDD+, PES, and NbS – reflects the same structural flaws and outlines what it would mean to reorient them around IPLC governance, negotiated legitimacy, and practical viability.

References

Acemoglu, D., & Robinson, J. A. (2012). *Why Nations Fail: The Origins of Power, Prosperity, and Poverty*. New York: Crown Business.

Adenle, A. A., et al. (2017). Managing climate change risks in Africa – A global perspective. *Ecological Economics*, 141, 190–201.

Africa Center for Strategic Studies. (2025). The regional security imperative to protect the Congo Basin. https://africacenter.org/publication/asb44en-regional-security-imperative-protect-congo-basin/

Allcott, H., & Mullainathan, S. (2010). Behavior and energy policy. *Science*, 327(5970), 1204–1205.

Aliansi Masyarakat Adat Nusantara (AMAN). (2021). *Submission to the United Nations Committee on Economic, Social and Cultural Rights (CESCR) on the Situation of Indigenous Peoples in Indonesia*. AMAN. Retrieved from https://www.aman.or.id/files/publication-documentation/9557Press%20Release%20AMAN%20-%20CESCR%20FINAL.pdf

APA Climate Change Task Force. (2022). *Addressing the Climate Crisis: An Action Plan for Psychologists*. American Psychological Association.

AP News. (2025, January 20). Trump signs order directing US withdrawal from the Paris climate agreement. https://apnews.com/article/trump-paris-agreement-climate-change-788907bb89fe307a964be757313cdfb0

Asiyanbi, A., & Massarella, K. (2020). Transformation is what you expect, models are what you get: REDD+ and models in conservation and development. *Journal of Political Ecology*, 27(1), 476–495. https://doi.org/10.2458/v27i1.23540

Bangalore, M.; Hallegatte, S.; Bonzanigo, L.; Kane, T.; Fay., M.; Narloch, U.; Treguer, D.; Rozenberg, J.; Vogt-Schilb, A.. 2016. Shock Waves: Managing the Impacts of Climate Change on Poverty. *Climate Change and Development*. © World Bank. http://hdl.handle.net/10986/22787, License: CC BY 3.0 IGO.

Bayrak, M., & Marafa, L. (2016). Ten years of REDD+: A critical review of the impact of REDD+ on forest-dependent communities. *Sustainability*, 8(7), 620.

Beaury, E. M., et al. (2024). Global potential and limits of land-based climate change mitigation strategies. *Nature Climate Change*, 14(1), 54–62.

Borras, S. M., & Franco, J. C. (2018). The challenge of locating land-based climate change mitigation and adaptation politics within a social justice perspective: Towards an idea of agrarian climate justice. *Third World Quarterly*, 39(7), 1308–1325.

Bristol Cabot Institute for the Environment. (2024). *New study highlights expansion of drylands amidst impact of climate change*. University of Bristol. Retrieved from https://www.bristol.ac.uk/cabot/news/2024/drylands-expansion.html

Brown, K., & MacLellan, L. (2020). Contesting carbon: The politics of measuring, valuing, and compensating forest carbon in policy and practice. *Global Environmental Change*, 64, 102123.

Brown, M. (2013). *Redeeming REDD: Policies, Incentives and Social Feasibility for Avoided Deforestation*. London and New York: Routledge. https://www.taylorfrancis.com/books/mono/10.4324/9780203123652/redeeming-redd-michael-brown

Brown, M., & Hughes, A. K. (2017, March). *Integrating Community-Specific Tenure Approaches in Rwanda: Lessons for Forest Landscape Restoration and Sustainable Forest Management*. Paper presented at the World Bank Land and Poverty Conference, Washington, DC.

Brown, M., Ngwala, P., Songo, A., Wande, L. (2004). Combating low-level corruption on waterways in the democratic republic of Congo: Approaches from Bandundu and Equateur provinces. *George Washington Law*. Washington, DC. Available at SSRN. https://ssrn.com/abstract=627684 or http://dx.doi.org/10.2139/ssrn.627684

Buchner, B., et al. (2021). *Global Landscape of Climate Finance 2021*. Climate Policy Initiative.

Büntgen, U. (2024). The importance of distinguishing climate science from climate activism. *npj Clim. Action*, 3, 36.

Büscher, B., & Fletcher, R. (2020). *The Conservation Revolution: Radical Ideas for Saving Nature Beyond the Anthropocene*. Verso.

Carlisle, J. E., et al. (2023). Understanding rural attitudes toward the environment and conservation in America. *Environmental Politics*, 32(2), 339–362.

Chaffin, B. C., Gosnell, H., & Cosens, B. A. (2014). A decade of adaptive governance scholarship: Synthesis and future directions. *Ecology and Society*, 19(3), 56.

Chausson, A., et al. (2020). Mapping the effectiveness of nature-based solutions for climate change adaptation. *Global Change Biology*, 26(11), 6134–6155.

Cheng, S., et al. (2023). The evidence base for the effectiveness of nature-based solutions to climate change impacts: A systematic map. *Environmental Evidence*, 12(1), 1–23.

CIFOR-ICRAF. (2025). National-level corruption risks and mitigation strategies in REDD+ implementation in the DRC. https://www.cifor-icraf.org/knowledge/publication/5552/

Climate and Land Use Alliance. (2015). Ghana REDD+ case study. https://www.climateandlandusealliance.org/reports/ghana-redd-case-study/

Cole, S., Ehret, P., et al. (2022). Are Americans really polarized on climate change? A meta-analysis. *Environmental Communication*, 16(6), 776–788.

Conservation International. (2024, October 29). *Six global environmental organizations unite to scale climate and conservation outcomes through sovereign debt conversions*. Retrieved from https://www.conservation.org/press-releases/2024/10/29/six-global-environmental-organizations-unite-to-scale-climate-and-conservation-outcomes-through-sovereign-debt-conversions

Convention on Biological Diversity (CBD). (2018). *Report of the Technical Expert Workshop on Other Effective Area-Based Conservation Measures for Achieving Aichi Biodiversity Target 11 (CBD/PA/EM/2018/1/2)*. Montreal, Canada: Secretariat of the Convention on Biological Diversity.

Council for Scientific and Industrial Research – Forestry Research Institute of Ghana (CSIR-FORIG). (2020). *Climate Change Adaptation and Forest Management in Ghana*. https://csir-forig.org.gh/portfolio-item/annual-report-2020/

Dawson, N. M., et al. (2024). Reviewing the science on 50 years of conservation: Knowledge production biases and lessons for practice. *Ambio*, 53, 1395–1413.

Druckman, J. N., & McGrath, M. C. (2019). The evidence for motivated reasoning in climate change preference formation. *Nature Climate Change*, 9(2), 111–119.

Dudine, P., & Szoke-Burke, S. (2020). Community participation and benefit sharing in Ghana's REDD+ program. *Columbia Center on Sustainable Investment*. https://ccsi.columbia.edu/sites/default/files/content/docs/publications/CCSI-REDD-Ghana-Report.pdf

Ecosystem Marketplace. (2023). *Paying for Quality: State of the Voluntary Carbon Markets 2023*. Forest Trends' Ecosystem Marketplace. https://3298623.fs1.hubspotusercontent-na1.net/hubfs/3298623/SOVCM%202023/2023-EcoMarketplace_SOVCM-Nov28_FINALrev-Mar2024.pdf

Ecosystem Marketplace. (2024). *2024 State of the Voluntary Carbon Market (SOVCM)*. https://www.ecosystemmarketplace.com/publications/2024-state-of-the-voluntary-carbon-markets-sovcm/

Edelman. (2020). Edelman Trust Barometer 2020. https://www.edelman.com/trust/2020-trust-barometer

Fagan, M., & Huang, C. (2019). *A Look at How People Around the World View Climate Change*. Pew Research Center.
Fanning, A. L., & Hickel, J. (2023). Compensation for atmospheric appropriation. *Nature Sustainability*, 6, 1077–1086.
Food and Agriculture Organization of the United Nations (FAO), CGIAR, & CARE International. (2021). *Deploying a humanitarian–development–peace nexus approach: Exploring, strengthening and reviving dryland ecosystems*. FAO. Retrieved from https://careclimatechange.org/wp-content/uploads/2021/09/FAO_CARE_drylandnexusreport.pdf
Forest Peoples Programme (FPP). (2021). *Position Statement: Indigenous Peoples, Local Communities, and 30x30 – Safeguarding Rights and Strengthening Conservation Outcomes*. Forest Peoples Programme.
Freund, J., (2025). From aid to incentives: Why forest conservation needs market solutions. https://africasolutionsmediahub.org/2025/05/20/from-aid-to-investments-why-forest-conservation-in-africa-needs-market-solution/
Gallup. (2021, December 16). Deconstructing Americans' views of socialism, capitalism. *Gallup News*. https://news.gallup.com/opinion/polling-matters/358178/deconstructing-americans-views-socialism-capitalism.aspx
Garnett, S. T., et al. (2018). A spatial overview of the global importance of Indigenous lands for conservation. *Nature Sustainability*, 1(7), 369–374.
Global Forest Coalition. (2024). Who Really Benefits? How REDD+ fails forests and those who protect them: A critical analysis of ongoing REDD+ implementation. https://globalforestcoalition.org/wp-content/uploads/2024/06/EN-REDD-paper-2024.pdf
Global Witness. (2019, December 9). Congo oil project obtained by "corruption risk" magnate threatens climate-critical peatland forest. Global Witness. https://www.globalwitness.org/en/press-releases/congo-oil-project-obtained-by-corruption-risk-magnate-threatens-climate-critical-peatland-forest/
Green-Win Project. (2018). Transformative narratives for climate action: Win-win strategies linking climate and sustainable development goals. Retrieved from https://www.green-win-project.eu/resource/transformative-narratives-climate-action-win-win-strategies-linking-climate-and-0
Gujo, M. M., & Modiba, L. M. (2025, January 9). Food insecurity confrontation by pastoralist and agrarian communities in South Omo Zone, Ethiopia: A facility-based qualitative study. *Journal of Nutritional Science*, 14, e1. https://doi.org/10.1017/jns.2024.88
Harwell, E. (2010). Forests in fragile and conflict-affected states. Program on Forests (PROFOR), World Bank. Retrieved from https://openknowledge.worldbank.org/server/api/core/bitstreams/79b95e9c-c3e6-56af-bc77-1fc10014c972/content
Haya, B., Buchanan, J., Berry, F., & Kelsey, H. (2023). *Quality Assessment of REDD+ Carbon Credit Projects*. Berkeley: Berkeley Carbon Trading Project, Goldman School of Public Policy, University of California.
Hornsey, M. J., & Fielding, K. S. (2020). Understanding (and reducing) inaction on climate change. *Social Issues and Policy Review*, 14(1), 3–35.
Hornsey, M. J., Harris, E. A., Bain, P. G., & Fielding, K. S. (2016). Meta-analyses of the determinants and outcomes of belief in climate change. *Nature Climate Change*, 6(6), 622–626.
IIED. (2020). Principles for locally led adaptation. https://www.iied.org/principles-for-locally-led-adaptation
Integrity Council for the Voluntary Carbon Market (ICVCM). (2023). *Core Carbon Principles, Assessment Framework and Assessment Procedure*. Integrity Council for the Voluntary Carbon Market.
International Institute for Environment and Development (IIED). (2024). Climate change and conflict – reframing the debate. https://www.iied.org/climate-change-conflict-reframing-debate.
International Union for Conservation of Nature (IUCN). (2020). *IUCN Global Standard for Nature-Based Solutions*. IUCN, International Union for Conservation of Nature.
International Union for Conservation of Nature (IUCN). (2023). *Africa Protected Areas Congress (APAC) Report*. IUCN.

Kabisch, N., et al. (2017). *Nature-Based Solutions to Climate Change Adaptation in Urban Areas: Linkages between Science, Policy and Practice*. Springer Nature.
Karsenty, A. (2014). Carbon concessions as a solution to deforestation? Lessons from the Congo Basin. In E. O. Sills, S. Atmadja, C. de Sassi, A. E. Duchelle, D. L. Kweka, I. A. P. Resosudarmo, & W. D. Sunderlin (Eds.), *REDD+ on the ground: A case book of subnational initiatives across the globe* (pp. 139–160). Center for International Forestry Research (CIFOR). https://doi.org/10.17528/cifor/005202
Klein, N. (2014). *This Changes Everything: Capitalism vs. the Climate*. Simon and Schuster.
Kyed, H. M., & Chambers, J. (2023). *Climate change actions in conflict affected contexts*. Danish Institute for International Studies (DIIS). Retrieved from https://www.diis.dk/en/research/climate-change-actions-in-conflict-affected-contexts
Lang, C. (2024, December 1). "The return of the carbon bomb: Democratic Republic of Congo opens 52 oil exploration blocks in the Cuvette Centrale peatlands." *REDD-Monitor (Substack)*. https://reddmonitor.substack.com/t/fossil-fuels
Legrand, T., & Gowae, G. (2020). Final evaluation report: Papua New Guinea REDD+ Readiness Project Phase 2. UNDP, Forest Carbon Partnership Facility, & World Bank. Retrieved from https://www.forestcarbonpartnership.org/system/files/documents/final_report_to_fcpf_2mar2021.pdf
Le Monde. (2024, November 13). COP29: Alone at the podium, developing countries call for action and funding. https://www.lemonde.fr/en/environment/article/2024/11/13/cop29-developing-countries-call-for-action-and-funding_6732664_114.html
Lemos, M. C., Arnott, J. C., Ardoin, N. M., Baja, K., Bednarek, A. T., Dewulf, A., Fieseler, C., Goodrich, K. A., Jagannathan, K., Klenk, N. L., et al. (2021). Who governs at what price? Technocratic dominance and adaptive choices in water governance. *Frontiers in Water*, 3, 735018.
Lincoln Institute of Land Policy. (2024). *Land-Based Carbon Mitigation Strategies: Challenges and Opportunities*. Policy Focus Report.
Lockwood, M. (2018). Right-wing populism and the climate change agenda: Exploring the linkages. *Environmental Politics*, 27(4), 712–732.
Lomborg, B. (2020). *False Alarm: How Climate Change Panic Costs us Trillions, Hurts the Poor, and Fails to Fix the Planet*. New York: Basic Books.
Monbiot, G. (2019, April 11). Comment in Frankie Boyle's new world order. *BBC Two*. https://www.youtube.com/watch?v=7wGscFxs6RY
Najam, A., Huq, S., & Sokona, Y. (2003). Climate negotiations beyond Kyoto: Developing countries' concerns and interests. *Climate Policy*, 3(3), 221–231. Retrieved from https://www.sciencedirect.com/science/article/abs/pii/S1469306203000573
Newell, P., & Taylor, O. (2018). Contested landscapes: The global political economy of climate-smart agriculture. *The Journal of Peasant Studies*, 45(1), 108–129.
Pacific Institute. (2022). *Stakeholder Engagement Guide for Nature-Based Solutions*. Pacific Institute.
Proforest Initiative. (2023). *Jurisdictional Approaches in the Forest, Agriculture and Commodity Trade (FACT) Dialogue: Lessons Learned to Date*. Proforest.
Radachowsky, J., Ramos, V. H., McNab, R., Baur, E. H., & Kazakov, N. (2012). Community forestry in the Maya Biosphere Reserve, Guatemala: Protection through participation. *Conservation and Society*, 10(1), 56–67. Retrieved from https://www.conservationandsociety.org
Rights and Resources Initiative (RRI). (2018). *A global baseline of carbon storage in collective lands: Indigenous and local community contributions to climate change mitigation*. Washington, DC: Rights and Resources Initiative. Retrieved from https://rightsandresources.org/wp-content/uploads/2018/09/A-Global-Baseline_RRI_Sept-2018.pdf
Rights and Resources Initiative (RRI). (2022). Community forest concessions in the Maya Forest: Lessons for climate and conservation. Accessed in: https://rightsandresources.org/publication/community-forest-concessions-maya-forest/
Rights and Resources Initiative. (2023). *Rights, climate and conservation: Annual report 2023*. Rights and Resources Initiative. Retrieved from https://rightsandresources.org/wp-content/uploads/RRI-Annual-Report-2023-WEB-EN.pdf

Roberts, A., & Lamp, N. (2021). *Six Faces of Globalization: Who Wins, Who Loses, and Why It Matters*. Cambridge, MA: Harvard University Press.
Roe, S., et al. (2021). Land-based measures to mitigate climate change: Potential and feasibility by country. *Global Change Biology*, 27(23), 6025–6058.
Scheidel, A., et al. (2020). Environmental conflicts and defenders: A global overview. *Global Environmental Change*, 63, 102104.
Schroeder, H. (2010). Agency in international climate negotiations: The case of Indigenous Peoples and avoided deforestation. *International Environmental Agreements: Politics, Law and Economics*, 10(4), 317–332.
Schulz, K. A., & Siriwardane R. (2015). Depoliticised and technocratic? Normativity and the politics of transformative adaptation. *Earth System Governance Working Paper No.33*. https://www.earthsystemgovernance.org/wp-content/uploads/2016/04/ESG-WorkingPaper-33_Schulz-and-Siriwardane.pdf
Seddon, N., et al. (2020). Understanding the value and limits of nature-based solutions to climate change and other global challenges. *Philosophical Transactions of the Royal Society B*, 375(1794), 20190120.
Seddon, N., et al. (2021). Getting the message right on nature-based solutions to climate change. *Global Change Biology*, 27(8), 1518–1546.
Sierra, J., & Suárez-Collado, Á. (2021). Understanding economic, social, and environmental sustainability challenges in the Global South. Sustainability, 13(13), 7201. https://doi.org/10.3390/su13137201
Souza, F., Tauli Corpuz, J., Sena, K., & Sucre Romero, L. (2023, November 27). How Indigenous peoples and local communities can make the voluntary carbon market work for them (commentary). Mongabay. https://news.mongabay.com/2023/11/how-indigenous-peoples-and-local-communities-can-make-the-voluntary-carbon-market-work-for-them-commentary/
Sovacool, B. K. (2021). Is sunshine the best disinfectant? Evaluating the global effectiveness of the Extractive Industries Transparency Initiative (EITI). *Energy Policy*, 156, 112622.
Sovacool, B. K., Baum, C. M., & Low, S. (2022). Climate protection or privilege? A whole systems justice milieu of twenty negative emissions and solar geoengineering technologies. *Political Geography*, 97, 102702. https://doi.org/10.1016/j.polgeo.2022.102702
Tauli-Corpuz, V., et al. (2020). Cornered by PAs: Adopting rights-based approaches to enable cost-effective conservation and climate action. *World Development*, 130, 104923.
Tengö, M., et al. (2017). Weaving knowledge systems in IPBES, CBD and beyond – Lessons learned for sustainability. *Current Opinion in Environmental Sustainability*, 26–27, 17–25.
Tesler, M. (2018). Elite domination of public doubts about climate change (not evolution). *Political Communication*, 35(2), 306–326.
Transparency International. (2021). *Corruption Perceptions Index 2020*.
UNDP Papua New Guinea. (2021). *REDD+ Readiness Phase 2: Final report*. Retrieved from https://www.forestcarbonpartnership.org/system/files/documents/final_report_to_fcpf_2mar2021.pdf
Van Boven, L., Ehret, P. J., & Sherman, D. K. (2018). Psychological barriers to bipartisan public support for climate policy. *Perspectives on Psychological Science*, 13(4), 492–507.
West, T. A., et al. (2020). Overstated carbon emission reductions from voluntary REDD+ projects in the Brazilian Amazon. *Proceedings of the National Academy of Sciences*, 117(39), 24188–24194.
World Bank. (2010). *World Development Report 2010: Development and Climate Change*. Washington, DC: World Bank.
World Bank. (2015). *World – Dedicated Grant Mechanism (DGM) for Indigenous Peoples and Local Communities including a Global Learning and Knowledge Exchange Project*. Washington, DC: World Bank Group. https://documents.worldbank.org/en/publication/documents-reports/documentdetail/616421468179664208
World Resources Institute. (2024). *Global South Climate Cooperation: Challenges and Opportunities*. WRI Report.
World Wide Fund for Nature (WWF). (2021). *Nature-Positive: WWF's Manifesto for a Nature-Positive World by 2030*. WWF International.

3 Nature-Based Solutions – Rhetoric, Reality, and Reform

The Rise, Rhetoric, and Contradictions of Nature-based Solutions

Nature-based Solutions (NbS) have rapidly ascended to become the dominant international strategy for land-based climate mitigation, particularly in the Global South. Defined by the International Union for Conservation of Nature (IUCN) as "actions to protect, sustainably manage, and restore natural and modified ecosystems that address societal challenges effectively and adaptively, simultaneously providing human well-being and biodiversity benefits" (Cohen-Shacham et al., 2016; UNEP, 2022).

This optimism is exemplified in recent European Commission policy, which elevates NbS to a central pillar of the European Green Deal and related initiatives:

> Nature-based solutions are a core element of the European Green Deal and recent major European policy initiatives, such as the EU Biodiversity Strategy for 2030 and the new EU Strategy on Adaptation to Climate Change. They play a key role in the new EU Forest Strategy, the overarching Water Framework Directive, and the European Zero Pollution Action Plan for air, water, and soil. Nature-based solutions also contribute to objectives established as part of international agreements, such as the Sendai Framework for Disaster Risk Reduction and the United Nations' Sustainable Development Goals. Through their role in the new European economy, nature-based solutions can also contribute to the targets established in the EU Bioeconomy Strategy and the EU Circular Economy Strategy, and in the "Fit for 55 package" on the transformation of the EU economy and society to meet climate ambitions. They are a key element of the transformative changes needed towards a sustainable future.
>
> (European Commission, Directorate-General for Research and Innovation, 2021)

Yet beneath this appealing rhetoric and policy elevation, the dominant model of NbS remains anchored in carbon metrics and the financialization of nature. The problem is not that NbS lack ecological potential, but that their implementation rarely prioritizes what makes interventions socially feasible or institutionally

DOI: 10.4324/9781003380320-4

appropriate. Too often, strategies are being deployed without secure land tenure, functional governance, sociocultural legitimacy, or capacity for co-design – especially in the Global South, where over 80% of land-based mitigation potential resides but governance is weakest (Roe et al., 2021).

A deeper contradiction emerges when examining participation and governance. Even as the European Commission and other major actors call for "collaborative approaches" and stakeholder involvement, these efforts frequently fall short. Participation is often limited to consultation or information provision, while key decisions remain top-down. The European Commision's "Golden Standard," for example, still guards decision-making prerogatives at the highest level, precluding true co-design or community authority (World Resources Institute, 2024):

> Monitoring and evaluation can benefit from collaborative approaches. Co-production is different from consultation or information provision in that stakeholders are involved from the beginning in the development of monitoring and evaluation plans. The knowledge, expertise, and lived experience of a wide range of stakeholders are relevant when deciding what outcomes to evaluate, identifying local needs, implementing monitoring strategies, and gathering relevant data. Using well-designed collaborative approaches can also reduce costs and enhance nature-based solutions ownership.
>
> (European Commission, Directorate-General for Research and Innovation, 2021)

Despite this language, most NbS programs still bypass community authority and impose frameworks that serve technocratic, carbon-centered agendas (Sobrevila, 2008). REDD+ and its successors illustrate this pattern: Carbon becomes the main metric of success, while tenure rights, biodiversity complexity, and livelihood resilience are sidelined. The result is not a failure of the NbS concept itself, but a credibility crisis rooted in a paradigm that overemphasizes quantification and overlooks feasibility (Beaury et al, 2024b; Dooley et al., 2022; Gvein et al., 2023; Reyes-garcia et al, 2022). As one corrective, attention to land tenure has grown, with a coalition of governments and private donors pledging $1.7 billion at COP26 to strengthen Indigenous and local communities' land rights and guardianship (Sutherland, 2021; Veit & Reytar, 2017; Walker et al., 2020; World Bank, 2022). The challenge, however, is not simply about follow-through on financial commitments.

What is needed now is a reset of priorities. NbS should not continue to be judged by carbon volumes or hectares conserved for financialization but by how well their STAM reflect community priorities, reinforce local governance, and help build long-term resilience. This means rooting mitigation in negotiated models, where social and technical feasibility is not an afterthought, but the precondition. This is the only way that credible financialization, negotiated between IPLCs, jurisdictional governments, and climate financiers, can be obtained. Only then can justice become more than an aspiration – it must be built on a foundation of appropriate, locally anchored practice (Allen, 2022; Rights and Resources Initiative, 2024), while supporting principles to achieve sustainability (Ostrom, 2009).

The current situation regarding NbS assumptions and rhetoric, evidence-based realities, and the proposed shift advanced in this book is summarized in Table 3.1.

This chapter critically examines the conceptual evolution, empirical performance, and governance implications of NbS in the Global South. Building on previous chapters' critiques of carbon-centered, technocratic climate models, it begins by unpacking the rhetorical appeal of NbS and interrogates its credibility by analyzing its reliance on carbon-first metrics, standardized MRV systems, and top-down implementation frameworks. REDD+ is treated as the flagship NbS initiative, serving as a test case of these structural flaws. The chapter concludes by advancing a shift toward an appropriate and negotiated NbS paradigm – one grounded in IPLC leadership, social license, and context-specific governance. Rather than discarding NbS entirely, it argues for strengthening the concept through a governance framework grounded in feasibility and appropriateness – secure tenure, local legitimacy, negotiated co-governance, and credible monitoring and adaptive management – through which real-world success and sustained functionality become possible (Sterling et al., 2017).

Carbon-First Frameworks: Metrics, Models, and Misalignment

The carbon-first approach to NbS emphasizes measurable outcomes. Yet, it fails to account for social, ecological, and governance complexities. MRV systems, while technically sophisticated, do not factor in tenure security, livelihood resilience, and IPLC involvement in more than a superficial way. This reinforces a framework that benefits external actors over local communities. This "mono-mission" culture – where organizations and funders chase a single, quantifiable output, carbon – has come at the expense of social acceptance, ecological integrity, long-term feasibility, and programmatic impact (BioCarbon Fund ISFL, 2024).

This pattern is not an aberration but a defining feature of the dominant NbS frameworks. Across the board, these frameworks continue to prioritize market access and compliance with MRV systems over social legitimacy or local fit. At scale, this technocratic focus has led to new forms of land conflict and enclosure, especially in places where tenure is insecure or land claims overlap (Huff & Naess, 2022; Snyder, 2023).

Table 3.1 Shift From NbS Rhetoric to Effective Action

NbS Rhetoric	*Reality on the Ground*	*Needed Shift*
Win-win solutions	Exclusion, land conflict, weak tenure	Negotiated, context-specific, not imposed
Carbon metrics	Social/ecological complexity ignored	Broader land-use indicators (STAM)
Participation claims	Top-down control persists	Genuine co-design, local authority

The tools that dominate the NbS space – carbon accounting, MRV systems, and remote sensing – are often promoted as objective and necessary for climate finance. In reality, they sideline tenure security, biodiversity, and the social outcomes that matter most to IPLCs. For example, REDD+ MRV frameworks define success by the volume of credits issued, not by whether communities benefit or ecosystems are managed sustainably. Even where safeguards are officially in place, they are inconsistently applied or enforced after the fact, undermining trust and IPLC rights (Dooley et al., 2022; Seddon et al., 2021).

The result is a system that erases the contributions of IPLCs, whose territories are often the most biodiverse and carbon-rich on the planet, while perpetuating what many now call "climate coloniality." External actors retain control over planning, data, and finance, while communities are left as recipients, not architects, of climate solutions (Dooley et al., 2022; Reyes-García et al., 2022).

Geospatial technologies like Geographic Information Systems (GIS) and remote sensing, now standard in NbS project design, can reinforce top-down control when not embedded in participatory governance. These tools have the potential to be powerful in the hands of IPLCs – supporting territorial defense and monitoring – but too often, they are used to map, measure, and manage landscapes without community consent or recognition of customary systems. The result frequently induces land conflict, ecological simplification, and the sidelining of local priorities (Ballesteros-Bolaños et al., 2024; Friess et al., 2016).

The financialization of NbS through carbon markets and PES exposes a further disconnect between market logic and rural realities. Revenues disproportionately accrue to intermediaries and consultants, while communities see marginal benefits and face new restrictions or even displacement. The permanence and integrity of these credits are increasingly questioned, as are the equity and justice of benefit-sharing arrangements (Badgley et al., 2022; Probst et al., 2024; Snilsveit et al., 2019). In many cases, the technical shortcomings of MRV and market-based models have contributed to a broader credibility crisis in voluntary carbon markets, where projects that overstate emissions reductions or fail to demonstrate "additionality" can still generate tradable credits, enabling polluters to claim carbon neutrality while continuing to emit (Probst et al., 2024; West et al., 2023).

The real issue is not a failure of the NbS concept itself, but a systemic problem in how these solutions are operationalized. Too often, strategies are deployed without the basic enabling conditions: secure land tenure, functional governance, sociocultural legitimacy, or the capacity for co-design (Roe et al., 2021). In short, NbS are rarely fit for purpose in the places where they are most needed. This overemphasis on quantification and market logic has led to a credibility crisis – not because NbS are conceptually flawed, but because they are embedded in a paradigm that overemphasizes quantification and overlooks feasibility (Dooley et al., 2022; Gvein et al., 2023; Reyes-García et al., 2022).

A reset of priorities is urgently needed. NbS should not be judged solely by carbon volumes or hectares conserved but by how well their STAM reflect community priorities, reinforce local governance, and build long-term resilience. This means rooting mitigation in negotiated models, where feasibility credible technical and social feasibility analysis is not an afterthought but a precondition. Only then can justice become more than an aspiration – it must be built on a foundation of appropriate, locally anchored practice (Allen, 2022; Rights and Resources Initiative, 2023).

Operationalization Through Tools and Markets

While the previous section has shown how the carbon-first paradigm shapes the overall logic of NbS, it is in the operational details – the tools, metrics, and market mechanisms – that these tendencies become most visible and consequential for communities and landscapes. The practical application of NbS depends on a suite of technologies and financial instruments that, while intended to deliver transparency and rigor, often reinforce the very disconnects between global ambition and local reality that undermine legitimacy and impact.

Carbon Accounting and MRV Systems

Carbon accounting and MRV systems are now the backbone of most NbS initiatives. These systems are designed to quantify greenhouse gas reductions in standardized units, enabling projects to participate in carbon markets and attract international finance. However, the technical sophistication of MRV can mask a narrow focus on emissions at the expense of broader social and ecological outcomes. For example, REDD+ MRV frameworks continue to define programmatic "success" by the volume of credits issued, rarely accounting for whether communities benefit or ecosystems are managed sustainably (Seddon et al., 2021). Even where safeguards are formally required, their application is inconsistent and often retroactive, which erodes trust and undermines the rights of IPLCs (Dooley et al., 2022).

This technical narrowing of what counts as success has real consequences. Tenure security, livelihood resilience, and Indigenous ecological knowledge are routinely left out of monitoring frameworks, despite being central to long-term stewardship (Chausson et al., 2020). The result is not just a gap in measurement but also a gap in legitimacy: Projects that overstate emissions reductions or fail to demonstrate "additionality" can still generate tradable credits, enabling polluters to claim carbon neutrality while continuing to emit (Probst et al., 2024). As long as MRV systems are designed without meaningful IPLC participation, they will continue to replicate both epistemic and material exclusion, undervaluing the contributions of Indigenous-managed lands even as evidence shows these territories are among the most biodiverse and carbon-rich on the planet (Garnett et al., 2018; Reytar & Veit, 2017).

GIS and Remote Sensing

GIS and remote sensing have become ubiquitous in the design and monitoring of NbS projects. These technologies enable large-scale mapping of carbon stocks and land-use changes and are often presented as neutral or objective tools. In practice, however, they can reinforce top-down control and ecological simplification when not embedded in participatory governance. For instance, in Thailand, afforestation suitability maps ignored customary land use, leading to mangrove restoration projects that displaced traditional fish nurseries and resulted in a 60% decline in local catches (Friess et al., 2016). In East Kalimantan, drones deployed under J-REDD enforced conservation rules without FPIC, escalating tensions with Dayak communities (Forest Peoples Programme, 2024).

Yet, these same technologies hold transformative potential when placed in the hands of IPLCs. Participatory geospatial platforms – such as QGIS and Google Earth Engine – are now being used for ethnoecological mapping, territorial defense, and the assertion of land rights. In the Amazon, the GeoIndigena platform combines satellite data with traditional knowledge to monitor illegal logging and defend Indigenous territories (Ballesteros-Bolaños et al., 2024). The legitimacy of GIS and remote sensing, then, derives not from technical precision alone but from their integration into local institutions and knowledge systems.

Carbon Markets and Offsetting

Carbon markets and offsetting mechanisms are central to the financialization of NbS, channeling billions of dollars into projects that promise measurable emissions reductions. Yet, the justice paradox of these markets is now well documented. Projects such as the Kasigau Corridor REDD+ in Kenya generated $50 million in credits, but less than 15% of revenues reached local communities, with the majority captured by intermediaries and technical consultants (WWF, 2013). The permanence of these credits is also questionable: Wildfires in California's forest offset projects have released up to 90% of sequestered carbon, exposing the fragility of market-based models (Probst et al., 2024) even in the United States.

Moreover, carbon markets have systematically privileged elite actors. In Africa, more than 60% of PES schemes disproportionately benefit large landowners, undermining smallholder participation and reinforcing existing inequalities (Snilsveit et al., 2019). The disconnect between market logic and on-the-ground realities is not incidental but structural, as it is a product of tools and frameworks that prioritize quantifiable outcomes over lived experience and local legitimacy.

Payment for Ecosystem Services

PES has become a widely adopted tool within the NbS paradigm, compensating landholders or communities for maintaining or enhancing ecosystem services such as carbon sequestration, watershed protection, or biodiversity conservation (Porras et al., 2013a; Wunder, 2005). In theory, PES aligns conservation incentives with local livelihoods. In practice, however, PES schemes often exacerbate inequity by failing to recognize customary land tenure, reinforcing unequal property rights, and in some cases facilitating land dispossession (Rights and Resources

Initiative, 2015a). In Kenya's Mau Forest, for example, PES programs allocated payments to titleholders, displacing Ogiek hunter-gatherers with ancestral claims (Marshall, 2023a). Metric rigidity is another challenge: Costa Rica's PSA program prioritized carbon and water metrics, excluding Indigenous agroforestry practices that enhance biodiversity.

The effectiveness of PES as a tool for achieving ecological and social outcomes remains contested. Its most notable successes have occurred where long-term institutional arrangements recognize Indigenous and local community rights and where governance systems are adaptive and participatory (Muradian et al., 2010). Without such frameworks, PES becomes a tool that extracts compliance from communities rather than empowering them to co-manage resources (Seddon et al., 2020). Systematic reviews show that PES schemes in low- and middle-income countries often deliver minimal poverty alleviation and fall short of their ecological goals (Snilsveit et al., 2019a).

Digital Innovations and Data Platforms

The latest wave of digital innovations – artificial intelligence, machine learning, blockchain, and participatory mapping platforms – promises to revolutionize NbS delivery and monitoring. AI-driven platforms such as Pachama use machine learning to verify forest carbon stocks, reducing MRV costs and increasing the speed of credit issuance. Yet, these proprietary algorithms often lack transparency and accountability, sidelining IPLC input and reinforcing "data extractivism" (Purwar & Akhter, 2023). Blockchain-based platforms like Carbon Credits tokenize offsets but cannot resolve the underlying issues of land rights or benefit-sharing.

Crucially, the legitimacy and effectiveness of digital tools depend on whether they are embedded in participatory governance frameworks. Platforms like explorer.land enable communities to visualize project impacts and assert territorial claims in NbS projects, but only a minority of NbS initiatives currently use such tools inclusively (Ballesteros-Bolaños et al., 2024). Digital innovation, in other words, is not a substitute for social license, tenure security, or negotiated co-governance. Without these foundations, new technologies risk deepening the very patterns of exclusion they claim to disrupt.

Natural Climate Solutions: Promise and Persistent Pitfalls

Natural Climate Solutions (NCS) have been widely promoted as a suite of ecosystem-based interventions – from mangrove restoration and peatland protection to forest regeneration – intended to sequester carbon while delivering biodiversity and livelihood co-benefits. The appeal is clear: Mangroves, for example, store three to five times more carbon per hectare than upland tropical forests, while also buffering coastlines and sustaining fisheries (IUCN, 2020). In theory, NCS should offer a win-win for climate, conservation, and communities.

Yet, as the preceding sections have shown, operational realities often fall far short of this promise. The carbon accounting frameworks that underpin most NCS projects are poorly suited to the complex land management practices of IPLCs. In Thailand, for instance, mangrove restoration projects, designed to maximize carbon sequestration, displaced traditional fish nurseries and led to a 60% decline in local catches, sparking widespread resistance (Friess et al., 2016). Elsewhere, forest regeneration initiatives have imposed restrictive conservation regimes without recognizing customary land-use rights or providing alternative livelihoods (Rights and Resources Initiative, 2024). These are not isolated incidents; they reflect a structural tendency within the NbS paradigm to prioritize external metrics over locally defined values (Allen, 2022; Rights and Resources Initiative, 2023).

Despite frequent commitments to co-benefits and project ownership, power over planning, data, and funding remains centralized in international institutions and technical intermediaries. Community participation is often limited to consultation or compliance, rather than genuine co-design or shared authority. As a result, the social and ecological trade-offs of NCS interventions are rarely negotiated transparently, and benefit-sharing arrangements remain opaque or exclusionary (Cheng et al., 2023). This gap between rhetoric and reality is particularly stark when it comes to economic development. While NbS proponents often tout job creation and income generation through ecotourism or PES, systematic reviews reveal that benefits are frequently captured by intermediaries or elites rather than by the communities who bear the costs (Cheng et al., 2023; Snilsveit et al., 2019a).

Carbon Markets and Structural Asymmetries

Structural Flaws in the Carbon Market Model

Carbon markets are not neutral instruments of climate policy; they are shaped by the economic and political forces of the actors who profit from them. From the outset, these markets have been structured to prioritize tradability, cost-efficiency, and investor confidence, rather than empowering local communities or strengthening governance systems (Lohmann, 2009; Papandreou, 2016). The design of these markets incentivizes the maximization of credit volume, often to the detriment of smaller, community-led initiatives (Mackenzie, 2016). For example, the costs associated with project design, validation, and monitoring – often exceeding $200,000 – make it difficult for local communities to engage unless they partner with commercial developers, who then control the project and the profits (Milne, 2005; VCMI, 2023).

The result is a system where financial intermediaries – certifiers, project developers, brokers, and Environmental, Social, and Governance (ESG) investors – capture the majority of the value generated, while the communities whose lands produce these credits see very little. An oversupply of cheap credits further drives prices down to as little as $0.55 per ton, leaving local stakeholders with razor-thin margins and minimal financial returns (Global Forest Coalition, 2023).

Under these pressures, project developers often cut corners on social safeguards. Community participation, FPIC, capacity building, and equitable benefit-sharing are treated as “overheads” rather than core components of project design. This undermines the integrity of the market, turning the promises of "co-benefits" into mere marketing tools rather than tangible, on-the-ground benefits for local communities (Carbon Market Watch, 2023).

Figure 3.1, generated from numerous public domain sources (see sources below), roughly illustrates the value of different stakeholders in the carbon market supply and value chains:

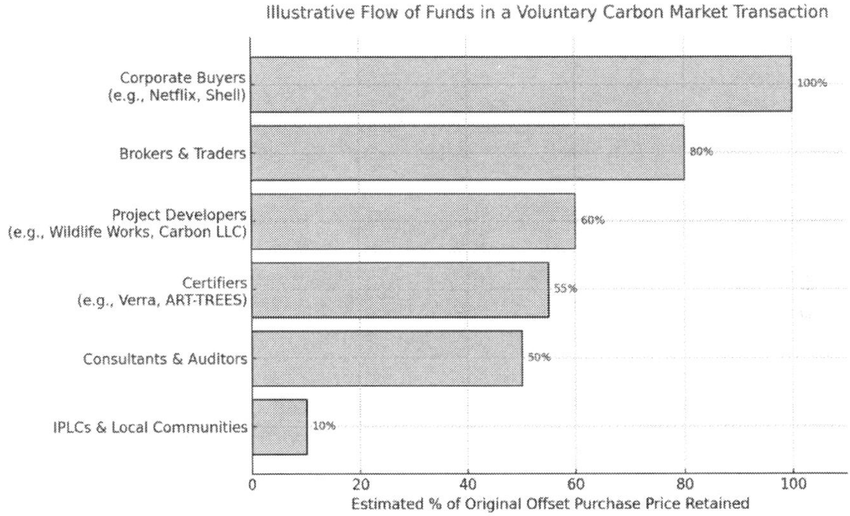

Figure 3.1 Illustrative Flow of Funds in a Voluntary Carbon Market Transaction

Sources: Based on synthesis of McKinsey & Company (2022), *A Blueprint for Scaling Voluntary Carbon Markets*; Carbon Market Watch (2023), *Ten Years of Carbon Market Scrutiny*; Forest Trends' *Ecosystem Marketplace* reports (2021, 2022, 2023, 2024); and case study investigations, including Kariba (The Guardian, 2023), Northern Rangelands Trust (Survival International, 2023) and Peru Shipibo REDD+ documentation (Forest Peoples Programme, 2023).

Under these pressures, project developers often cut corners on social safeguards to stay viable. Meaningful community participation, FPIC, capacity building, and equitable benefit-sharing are treated as costly "overheads" rather than as core components of a project (Papandreou, 2016). When a project's success is judged mainly by low transaction costs and high credit volumes, there is a temptation to sideline governance reform, legal support, and direct community benefits – the very things that ensure legitimacy.

Crucially, this is not a case of a few bad projects or technical errors; it is a structural feature of the carbon market's political economy (Lohmann, 2009). The system, by design, rewards volume and visibility over equity and justice, systematically marginalizing the IPLCs who are supposedly at the heart of NbS (Garnett et al., 2018).

In practice, IPLCs are frequently treated as peripheral stakeholders rather than decision-makers: Participation is reduced to a token compliance step, technical tools (GIS, carbon models) are deployed without local input, and success is measured in carbon tons rather than in tenure security or community well-being (Reytar & Veit, 2017). This top-down model concentrates authority in external institutions and experts, while the people most critical to mitigation – those living on and caring for the land – have little say. Unless the fundamental incentive structures (e.g., certifier fee models, project design norms, benefit-sharing arrangements) are overhauled, carbon markets will continue to privilege carbon accounting and investor returns at the expense of local rights and realities (McKinsey & Company, 2022).

Evidence From Carbon Offset Projects

Real-world carbon offset projects around the world illustrate these asymmetries between global market actors and local communities. For example:

- Kasigau Corridor REDD+ (Kenya): This flagship project generated about \$50 million in carbon credit revenues, yet less than 15% of that money reached the local communities (Kill, 2016).
- Kariba REDD+ (Zimbabwe): Touted as one of the largest forest conservation offset projects, Kariba sold tens of millions of dollars' worth of credits. However, it has been widely criticized for minimal community benefits and questionable carbon baselines. Independent investigations found that most of the project's €100 million (≈\$110 million) in revenue never reached the villages involved – only about 14% of funds were distributed to local people, with the rest absorbed by the developer and brokers (Carbon Market Watch, 2023; The Guardian, 2023). Kariba's case underscores how even "successful" projects can fail to deliver meaningful benefits on the ground while possibly overstating their climate impact.
- Northern Rangelands Trust (NRT): The Northern Kenya Rangelands Carbon Project enrolled pastoral communities without full FPIC or transparent contracts, and then carbon credits were sold to global companies like Netflix and Meta. Local residents reported little knowledge of where the carbon money went (Kimeu, 2025; Mukpo, 2025; Survival International, 2023). This exemplifies how deals are often struck by project proponents and buyers over the heads of communities, who are left with new land-use restrictions, murky projected benefits, and thoroughly disrupted customary land-use systems. In May 2025, Verra suspended the project after years of community pushback and backing from Kenya's courts.
- Trees for Global Benefit (Uganda): A community carbon forestry program in Uganda, intended to pay small farmers for tree-planting, ended up exacerbating food insecurity. Smallholders shifted land into trees but received only token payments, while the bulk of profits came from selling the credits to European buyers (D.K. & Kamukama, 2022). Weak land rights and power imbalances meant farmers bore project risks (and lost farmland for food) but saw little reward, reflecting North–South inequalities in carbon finance, raising questions about claims of climate benefit from these projects (The Guardian, 2024).

- Shipibo and Kuy Indigenous Territories (Peru and Cambodia): In forest carbon initiatives on Indigenous lands in Latin America and Southeast Asia, contracts were signed under highly asymmetrical information and pressure, offering communities limited understanding or say in the terms (Forest Peoples Programme, 2023; Global Witness, 2024). Such cases show how IPLCs can be pushed into agreements that they do not fully consent to, in exchange for uncertain future benefits, effectively continuing a colonial pattern of resource control under the banner of carbon conservation.

These examples are far from isolated outliers; they are symptomatic of the current system. A recent analysis by McKinsey concluded that, on average, only 10%–20% of the value of carbon offset credits ultimately reaches local communities, with the remaining 80%–90% captured by project developers, certifiers, consultants, and brokers (McKinsey & Company, 2022). This skewed distribution of benefits is evidence of a structural imbalance in carbon markets – one that consistently favors outside actors financially and leaves the supposed beneficiaries with a marginal share. This also endangers future NbS projects (see Box 3.1).

Box 3.1 Perceived Successes and Hidden Challenges in IPLC-Led NbS Mitigation

While some Nature-based Solutions (NbS) projects (IUCN, 2022) involving Indigenous Peoples and Local Communities (IPLCs) have achieved measurable climate mitigation outcomes, the term "successful" requires nuance. Success in these contexts often hinges on several interconnected factors, including secure land rights, community governance, and equitable benefit-sharing. In cases where IPLCs have both legal land tenure and active governance over their lands, positive outcomes in carbon sequestration, biodiversity protection, and climate resilience are more likely to occur. For instance, Mexico's community forestry initiatives and Canada's Indigenous-led forest management have demonstrated that secure land rights enable IPLCs to apply traditional ecological knowledge alongside modern conservation methods, fostering long-term environmental and climate benefits (Berkes, 2007; Bray et al., 2003).

However, even when land rights are secure, challenges remain that can complicate the perceived success of these projects. The Makira Forest project in Madagascar, for example, highlights that legal land tenure alone is not sufficient to ensure community success. Despite early claims of forest conservation success, the lack of negotiated benefit-sharing led to discontent and pushback from local communities. Financial and social benefits were not adequately distributed, leading to community frustration (Wiggins et al., 2018). Similarly, in the Xingu River Basin project in Brazil, despite strong local leadership and community participation, tensions arose from inequitable resource allocation and a lack of transparency in decision-making processes, which undermined long-term project sustainability (Marten et al., 2019).

In these cases, the real challenge lies not only in securing land rights but also in ensuring that the governance frameworks incorporate transparent, equitable benefit-sharing mechanisms. Without these elements, even well-structured projects can face opposition and may fail to deliver on their climate mitigation and community welfare promises (Ostrom, 1990).

Reclaiming Governance: Centering IPLC Authority

If carbon markets and NbS are to genuinely deliver climate benefits and community development, they must be radically restructured around principles of appropriateness, feasibility, and shared governance. In practice, this means moving away from an offset-centric, volume-driven paradigm toward one that prioritizes local authority and social legitimacy. The examples cited in Box 3.1 show that even with formal land tenure, the absence of effective governance and participatory safeguards can undermine conservation and development outcomes. International standards increasingly recognize the necessity of obtaining FPIC from IPLCs as a fundamental practice for ensuring genuine and equitable participation (ISEAL Alliance, 2020).

The legitimacy of any climate mitigation project should be defined not by how quickly it can issue carbon credits, but by whether the people who live in and steward the target landscapes agree that the project is appropriate and beneficial (Garnett et al., 2018). In other words, success must be measured in terms of community consent, land tenure security, and long-term sustainability, rather than just tons of CO_2 quantified.

A carbon project that meets all the technical standards is meaningless if it lacks the support and trust of the Indigenous and local communities on the ground (Reytar & Veit, 2017). Thus, the metric of "success" needs to shift from carbon efficiency to community approval and well-being. A key step is to turn the very tools of carbon accounting into instruments of empowerment for IPLCs. Technologies like satellite monitoring, GIS mapping, and carbon modeling have the potential to support community land management – but only when used by and with local communities rather than imposed on them (Ballesteros-Bolaños et al., 2024).

For example, when Indigenous organizations are trained and equipped to use mapping platforms or drone imagery, they can document their territories, monitor forest conditions, and detect illegal activities on their own terms (Ballesteros-Bolaños et al., 2024). In such cases, high-tech tools become means of asserting land rights and co-producing knowledge, instead of tools of external surveillance.

The overarching goal is to democratize data and planning: Rather than experts remotely deciding how a forest should be managed, communities should be co-designing interventions, with technology serving their strategies. Simply put, carbon MRV (monitoring, reporting, and verification) should be reimagined as participatory and community-driven so that the people most affected have control over how data is gathered and used.

Achieving this vision requires a shift from one-off, project-level consultation toward institutionalized co-governance at landscape and jurisdictional scales. IPLCs need to be recognized not just as stakeholders but as decision-makers with authority over project design and implementation (Reytar & Veit, 2017).

FPIC should be embedded as an ongoing governance principle – meaning communities have the right to veto, halt, or redesign activities at any stage if those activities violate agreements or local values (Garnett et al., 2018). Mechanisms for joint decision-making can include community assemblies or councils that hold real power in approving project plans, benefit-sharing arrangements that are negotiated and legally enforced, and grievance redress systems that are accessible to local people, whose priorities and interests are served (Mace, 2014).

Notably, such transformative approaches are already being proposed and piloted outside the dominant carbon-market framework. One example is the Tropical Forests Forever Facility (TFFF), envisioned as a performance-based funding platform to reward countries (and communities) for preserving intact forests without creating tradable offset credits. Instead of funneling money through carbon project developers, TFFF would disburse finance directly to national or local institutions based on independently verified conservation outcomes (Ballesteros-Bolaños et al., 2024).

In this model, a forest-rich country or region that maintains or increases its forest cover (and thus carbon stocks) would receive payments as a form of climate service reward, but these payments are not sold as offsets to polluters abroad. This approach aims to bypass the broker-heavy, credit-by-credit market in favor of funding arrangements that channel resources straight to governments and IPLC organizations for agreed-upon climate results (Ballesteros-Bolaños et al., 2024). Similar thinking underlies proposals for negotiated governance platforms described in later chapters, where governments, Indigenous/local representatives, and private investors co-manage climate initiatives. In this setup, all parties would negotiate the terms of land-use interventions, responsibilities, and benefit-sharing before any money moves.

Crucially, the private sector's role would shift from buying cheap offsets to partnering in equitable, regulated land-use agreements that uphold community rights and provide verified climate benefits. Carbon sequestration or offsetting would be a product of holistic land management efforts, not the primary focus.

These emerging ideas demonstrate that it is possible to support climate mitigation without replicating the extractive dynamics of current offset markets. Ultimately, what is needed is not "better" carbon offsets but a new architecture for climate finance – one that treats social license, consent, and equity as non-negotiable foundations of success (Reytar & Veit, 2017).

In this reimagined framework, IPLCs would be recognized as the architects and co-authors of climate solutions on their lands, rather than as passive beneficiaries of outside projects. Climate initiatives would thus be built upon local knowledge systems, cultural values, and community-defined priorities, with external support aligning with these, not the other way around. Without such a paradigm shift, the promise of nature-based climate solutions will remain severely compromised.

As observers have noted, the current system will continue to reward volume over value, optics over outcomes, and certification checkboxes over genuine consent (Papandreou, 2016). The future of effective land-based mitigation therefore hinges on restructuring power and governance: placing control back into the hands of the communities who know their environments best and have the greatest stake in sustaining them. Only by correcting the structural asymmetries in carbon markets – and elevating IPLCs from marginal participants to true partners – can climate action and justice be aligned to achieve lasting results on the ground (Papandreou, 2016).

Nationally Adapted Approaches for NbS and NCS

If the past decade has proven anything about NbS and NCS, it is that standardized, top-down models rarely succeed. Effective NbS and NCS must be tailored to the unique ecological, social, and governance conditions of each country and community and built with – not just for – the people who live there. Evidence from the Global South repeatedly demonstrates that participatory, locally adapted models, which embed local knowledge, emphasize co-design, and invest in capacity building, deliver stronger and more durable climate and social outcomes (Chausson et al., 2020; Wolff et al., 2022).

Programs that prioritize adaptation and participation offer important lessons. In Fiji and Indonesia, the RISE program put residents at the center of flood adaptation, training community members to collect their own data and shape project design. This approach required more time and resources but resulted in solutions that fit local needs and fostered genuine ownership (Wolff et al., 2022).

Similarly, Costa Rica's PES program worked best where Indigenous land rights were recognized and community-driven planning was supported but reinforced inequalities or even led to displacement where these conditions were absent. These cases highlight that success depends on strong local institutions, clear land tenure, and authentic participation from the outset (Chausson et al., 2020).

Despite these successes, most NbS and NCS projects still struggle to break from entrenched patterns. Governments and donors often default to "grey" infrastructure – such as dams, levees, or monoculture plantations – because these options are politically safer, easier to monitor, and more familiar to planners (Kabisch et al., 2022). Institutional fragmentation and sectoral silos further reinforce this path dependency: Ministries and agencies typically work in isolation, each with separate budgets and priorities, making it difficult to coordinate across sectors and design integrated solutions (Dunlop et al., 2024; Kabisch et al., 2022).

Bureaucratic inertia compounds these challenges, resulting in policies that do not align and missed opportunities for synergy. Monitoring and reporting requirements can become burdensome, often burying the real needs of communities under layers of paperwork (Seddon et al., 2020).

Overcoming these hurdles requires deliberate investment in participation and flexibility at every level. Establishing national advisory councils or stakeholder platforms where Indigenous Peoples, local communities, women, and youth have genuine decision-making power is essential. Such bodies can guide project design, implementation, and monitoring, ensuring that diverse perspectives are integrated from the start. Investing in the skills, knowledge, and resources of communities is critical for effective participation. Capacity-building measures must be tailored to local socio-economic contexts and power dynamics and should provide ongoing support beyond one-off training to ensure equitable benefit sharing in NbS and NCS activities (Forest & Climate Leaders Partnership, 2024).

Breaking down sectoral walls – so that climate, agriculture, water, and development teams work together – is crucial for designing solutions that are both effective and legitimate (Akimowicz et al., 2023). This requires changes in internal cooperation, working structures, and capacities to shift from siloed approaches to intersectoral work.

While technologies such as Artificial Intelligence, remote sensing, and satellite-based monitoring can optimize NbS delivery and improve transparency, their deployment must be guided by local leadership and ethical frameworks. Over-reliance on technology risks reinforcing data extractivism and replicating top-down governance failures. Ensuring data sovereignty, participatory data governance, and recognition of Indigenous and local knowledge is essential for just and effective outcomes (Allen, 2022; Purwar & Akhter, 2023; Rights and Resources Initiative, 2023).

The evidence is clear: Participatory, locally adapted NbS and NCS models deliver stronger climate and social outcomes. Achieving this shift demands political will, resources, and a willingness to move beyond top-down control. It is not easy, but it is the only way these solutions will endure and deliver on their promise (Chausson et al., 2020; Wolff et al., 2022). The effectiveness and equity of conservation fundamentally depend on centering IPLCs in design and implementation (Dawson et al., 2021).

Critical Assessment of Major NbS Instruments: REDD+, NCS, AFOLU, PES, and J-REDD

Earlier in this chapter, the challenges facing NbS were identified. As argued, the suite of STAM deployed by NbS – such as carbon accounting and MRV systems, GIS and remote sensing, carbon markets and offsetting, PES, REDD+, and emerging digital platforms – are not neutral. Instead, they reflect and reinforce a paradigm that privileges quantification, external design, and oversight, and market logics over the social, cultural, and ecological realities of communities and landscapes in the Global South. Each of these NbS "solutions," while defensible in theory, often fails in practice to capture the complexity of tenure, governance, and local knowledge that underpins durable climate action (Girardin et al., 2021; Seddon et al., 2020). What is troubling is that few are challenging the underlying premises of the STAM being deployed in NbS, with the European Commission Handbook cited earlier as a prime example of why current practice stands little chance of being successful, based on the totality of the publicly available evidence base.

Agriculture, Forestry, and Other Land Use

Agriculture, Forestry, and Other Land Use (AFOLU) is one of the most widely adopted land-based mitigation categories within the NbS paradigm. It encompasses strategies such as agroforestry, sustainable intensification, improved forest management, and afforestation/reforestation (A/R). AFOLU is embedded in over 90% of the Nationally Determined Contributions (NDCs) of Global South countries, underscoring its perceived centrality to climate mitigation efforts (IPCC, 2022; Roe et al., 2021). In theory, the framework offers a holistic platform to link food systems, forest conservation, and climate goals while supporting rural livelihoods.

Yet in practice, AFOLU reproduces many of the technocratic assumptions and governance failures found across the broader NbS paradigm. Its strategies often prioritize emissions reduction over social feasibility, overlooking tenure insecurity, customary land-use systems, and competing rural livelihood needs (Dooley et al., 2022; Roe et al., 2021). Tools such as carbon accounting and land suitability mapping are deployed without sufficient regard for historical exclusion or ecological complexity (Seddon et al., 2020).

Despite its widespread inclusion in NDCs, only 18% of AFOLU-related actions include concrete plans for IPLC engagement (Roe et al., 2021). The Land Gap Report highlights that land earmarked for mitigation often far exceeds what is socially or ecologically viable, particularly when IPLC land claims are considered (Dooley et al., 2022). In the Democratic Republic of the Congo (DRC), elites diverted 60% of REDD+ and AFOLU funds intended for communities, exposing acute governance failures (Kengoum et al., 2022). These are not isolated issues but structural patterns that continue to undermine legitimacy. As a result, AFOLU remains a paradox: expansive in potential but narrow in practice. Without participatory governance and equitable financing, it risks replicating the very limitations it seeks to solve.

Afforestation/Reforestation

As a core component of many AFOLU strategies, afforestation and reforestation (A/R) initiatives are often touted for their clear metrics – hectares planted, carbon sequestered, and alignment with NDCs. Their appeal lies in the ease with which they can be monitored, reported, and verified within carbon accounting systems (IPCC, 2022). However, this apparent simplicity often conceals ecological and social complexities that undermine long-term sustainability. Globally, up to 40%

of land targeted for A/R overlaps with areas already used for agriculture, raising serious concerns about food security and land conflict (Beaury et al., 2024a). From a STAM perspective, A/R illustrates how strategies focused on carbon often override social feasibility and ecological appropriateness. Technocratic tools like GIS-based suitability maps compromise Indigenous knowledge and local governance dynamics. Community members are frequently treated as labor inputs rather than partners in decision-making, and project methods emphasize planting targets over long-term co-management or benefit-sharing (Reyes-García et al., 2022; Tauli-Corpuz et al., 2020). Without participatory planning, tenure recognition, and adaptive governance, A/R risks replicating past models of enclosure and displacement under the guise of climate restoration. Its future depends on a paradigm shift: from extractive carbon logic to regenerative, place-based land governance that centers IPLC knowledge, autonomy, and cultural landscapes (Scheidel & Work, 2018).

REDD+ as Flagship Failure

Launched in 2005 under the UNFCCC, REDD+ is a global initiative aimed at incentivizing forest conservation in developing countries by offering payments for verified emissions reductions from avoided deforestation and degradation. By attaching market value to standing forests, REDD+ seeks to internalize environmental costs typically ignored in land-use decisions. Its scalability has made it a cornerstone of many mitigation portfolios. As of 2022, REDD+ programs covered more than 45 million hectares, and the IPCC estimated that under optimal conditions, they could contribute to reducing between 5% and 30% of global emissions (IPCC, 2022). Over 85% of tropical forest countries have integrated REDD+ into their NDCs, reflecting its alignment with global climate policy priorities (CIFOR-ICRAF, 2024). In Brazil, the need for integrated governance that addresses land tenure in both public and private lands is imperative (da Silva et al., 2023).

Despite its original theoretical promise, REDD+ has struggled to meet expectations over the past 20 years. One of its core structural flaws remains its overreliance on carbon accounting and offset markets, which has diverted attention from local governance, land tenure, and livelihood concerns reliant on the logic of integrated land use management systems. While learning from early REDD+ projects occurred (Massarella et al., 2018), there is no clear evidence this has led to replicated success as of 2025. REDD+ credibility has further eroded with evidence that more than 80% of its carbon credits lack "additionality," enabling polluters like Shell to maintain emissions while claiming neutrality (Probst et al., 2024), substantiating greenwashing concerns.

These critiques are not new. REDD+ has long been described as a "colonial mechanism" that commodifies Indigenous territories under the guise of mitigation (Cabello & Gilbertson, 2012; Waweru, 2016). Its governance structures have disproportionately favored large international NGOs and carbon intermediaries, sidelining IPLC actors whose stewardship is critical to forest sustainability (Green Climate Fund, 2024). In the DRC, centralized REDD+ implementation has enabled elite capture and reinforced exclusionary land politics (Kengoum et al., 2022).

Community-Managed Forest Systems: Ejidos as a Model for Nature-based Solutions

In the context of NbS, ejidos in Mexico represent a significant example of community-managed forest systems that align with key NbS principles. Ejidos are collective landholdings typically shared between Indigenous and local communities, where sustainable forest management is a central practice. These systems embody various NbS attributes such as the following:

- Sustainable forest management focused on carbon sequestration, soil conservation, and biodiversity protection. Ejidos have often been recognized for their contributions to maintaining healthy ecosystems and storing carbon, which mitigates climate change (Bray et al., 2003).
- Agroforestry systems that integrate tree planting with agricultural practices, creating landscapes that enhance ecosystem services, improve soil fertility and boost resilience to climate impacts. The combination of trees and crops in these systems helps to balance economic needs with environmental stewardship (López & Shapiro, 2018).
- Community participation is a cornerstone of ejidos, where local governance structures and communal decision-making processes are integral to forest management outcomes. In this model, land tenure security is directly tied to community empowerment, enabling local people to make decisions about their land use that also align with environmental conservation goals (Alatorre & Frenk, 2000; Bee, 2019).

Further, ejidos have successfully collaborated with both the state and private sector, engaging in certification programs like the Forest Stewardship Council (FSC) and participating in PES mechanisms. These collaborations enhance the economic sustainability of community-managed forests, providing market access while ensuring that the benefits of forest management are shared equitably within the community (Bray et al., 2003).

However, while ejidos provide a strong example of community-managed NbS, their success is not without challenges. Limited funding, fluctuating market conditions, and external pressures such as land-use changes and forest degradation pose ongoing risks. The experience of Mexico's ejidos highlights the critical importance of land rights in enabling local communities to succeed in NbS strategies and stresses the need for continued support to strengthen community governance and sustainable practices (Bee, 2019; Bray et al., 2003).

Payment for Ecosystem Services

PES has emerged as a widely adopted mechanism within the NbS paradigm. PES compensates landholders or communities for maintaining or enhancing ecosystem services, such as carbon sequestration, watershed protection, or biodiversity conservation. Proponents argue that this mechanism aligns conservation incentives with local livelihoods, fostering win-win solutions (Le at al., 2023; Porras et al., 2013; Wunder, 2005).

However, as broader critiques of NbS illustrate, PES exemplifies a reductionist approach to land management – isolating ecosystem functions for market exchange without considering the socio-political realities of land tenure and governance. In practice, PES has often exacerbated inequity. Though PES schemes promise mutually beneficial outcomes, they frequently fail to acknowledge customary land tenure, thus reinforcing unequal property rights and, in some instances, facilitating land dispossession (Rights and Resources Initiative, 2015a).

For example, in East Africa, many PES programs have disproportionately benefited elite landowners, sidelining marginalized groups and undermining traditional stewardship practices (Snilsveit et al., 2019). The appeal of PES, in theory, doesn't always translate into real-world success. Evidence shows that, when implemented without careful attention to local governance structures, PES can further entrench existing power imbalances, especially in low- and middle-income countries.

Nonetheless, some notable successes of PES have emerged in contexts where long-term institutional arrangements recognize the land rights of IPLCs, and where social capital is high on collectively managed lands (Kaiser et al., 2023). These cases tend to occur where governance systems are adaptive, participatory, and based on established community rights frameworks. A strong example of this is seen in Mexico's ejidos, discussed earlier, where the formal recognition of communal land rights has been integral to the success of forest management programs. These communities not only manage large forest areas but also collaborate with both the state and the private sector through mechanisms like the FSC certification, showcasing how local rights and governance can empower successful conservation models (Bray et al., 2008).

However, without such robust governance frameworks, PES often functions as a tool for extracting compliance from communities, rather than empowering them to co-manage resources. Systematic reviews reveal that PES schemes in low- and middle-income countries often deliver minimal poverty alleviation and fall short of achieving their ecological goals (Snilsveit et al., 2019). This misalignment between stated objectives and lived realities highlights a core challenge in NbS: Such frameworks cannot achieve legitimacy unless they center the governance rights, cultural frameworks, and material needs of the communities who sustain these landscapes.

Natural Climate Solutions

NCS encompass a suite of ecosystem-based interventions, such as mangrove restoration, peatland protection, and forest regeneration, intended to sequester carbon while delivering biodiversity and livelihood co-benefits. Advocates argue that these strategies harness nature's inherent capacity for climate regulation and offer synergies between conservation and development. For example, mangrove ecosystems can store three to five times more carbon per hectare than upland tropical forests while also buffering coastlines and sustaining fisheries (IUCN, 2020).

Yet despite their ecological potential, large-scale NCS initiatives frequently obscure the social trade-offs they generate, particularly for IPLCs. As Beaury et al. (2024a) argue, land-use suitability assessments for mitigation often fail to navigate competing claims between agricultural production, conservation, and carbon sequestration (Beaury et al., 2024a). These tensions are rarely negotiated transparently, leading to exclusionary project design and opaque benefit-sharing arrangements.

Although framed as ecologically neutral, many NCS programs reproduce the same extractive dynamics that have plagued REDD+: technocratic planning, elite brokerage, and the marginalization of IPLC land governance. Despite the inclusive language of NCS, studies show that community participation in agenda-setting, financing, or benefit allocation is rare (Cheng et al., 2023; Dooley et al., 2022). As a result, the legitimacy of these interventions – both social and ecological – remains contested.

Many projects unfold in what Huff and Naess (2022) describe as "expanding sacrifice zones," where IPLCs are expected to bear environmental burdens while global actors reap the rewards (Allen, 2022; Huff & Naess, 2022). Meanwhile, technical intermediaries, carbon brokers, and Global North institutions retain control over metrics, markets, and finance (Allen, 2022; Reyes-García et al., 2022).

Sustainable and Regenerative Agriculture

Sustainable and regenerative agriculture represent some of the most grounded and socially feasible strategies within the NbS paradigm, especially for IPLCs. Unlike top-down carbon offset schemes, these approaches emerge from longstanding land stewardship practices that emphasize reciprocity, food sovereignty, and ecological care.

Sustainable agriculture refers to methods that meet present food needs without compromising future generations' ability to do the same. It promotes biodiversity conservation, resource efficiency, and economic viability through techniques such as crop rotation, integrated pest management, water conservation, and soil fertility enhancement (FAO, 2021; SGS, 2024). These practices support climate resilience while sustaining rural livelihoods – objectives that align closely with the stated goals of NbS.

Regenerative agriculture builds on the foundation of sustainable farming by explicitly aiming to restore degraded ecosystems through practices such as no-till farming, cover cropping, composting, and rotational grazing. In the aggregate, the interventions increase soil organic matter, enhance water retention, support biodiversity, and offer meaningful climate mitigation co-benefits by sequestering carbon and reducing vulnerability to extreme weather events (EIT Food, 2023; General Mills, 2023).

The concept of regenerative agriculture was originally developed and championed by the Rodale Institute in the 1980s, well before carbon markets or NbS frameworks gained traction. Rodale's early field trials demonstrated that biologically rich soils could outperform conventional systems in carbon sequestration and drought resilience, framing healthy soils as central to sustainable food systems (Rodale Institute, 2020).

While this vision emphasized ecological complexity and farmer agency, much of today's regenerative discourse has been subsumed into technocratic models narrowly focused on carbon yield, sidelining the broader agroecological values at the heart of Rodale's original paradigm. Though much of the regenerative discourse has originated in the Global North, its principles are increasingly relevant in the Global South. In Kenya, India, and Brazil, smallholders are adapting agroecological models that sustain livelihoods and restore ecosystems, often without the need for carbon metrics or expensive technical interventions.

Regenerative agriculture challenges the logic of carbon commodification by highlighting culturally embedded knowledge systems and adaptive land governance. Yet even these interventions cannot be prematurely labeled "solutions." As with all NbS strategies, the legitimacy of sustainable and regenerative agriculture depends on demonstrable outcomes – measured not only in carbon or yield but also in equity, agency, and resilience (EIT Food, 2023; General Mills, 2023).

By embedding co-creation, cultural relevance, and long-term soil care into NbS portfolios, sustainable and regenerative agriculture potentially offer the kind of holistic, inclusive interventions that current frameworks lack, though scaling these represents a major challenge to maximize climate mitigation impacts when considered globally (Sustainable Markets Initiative Agribusiness Task Force, 2025).

Jurisdictional REDD: Hope and Limits

Jurisdictional REDD+ (J-REDD) has been promoted as a solution to the fragmentation and inefficiencies of project-based REDD+. Rather than relying on isolated initiatives led by external actors, J-REDD frameworks aim to embed emissions reduction within national or subnational governance systems at the provincial, regional, or landscape level. This approach is intended to enable greater policy coherence and alignment between climate goals and land governance reforms, with the potential, if properly structured, to support more participatory, territorially grounded, and equitable mitigation outcomes (Dooley et al., 2022a; Duchelle et al., 2018).

Some observers argue that J-REDD schemes may avoid the distortions and fragmentation associated with project-based REDD+, particularly when embedded in broader land governance reform and focused on landscape-level benefits beyond carbon (Boyd et al., 2018; Seymour & Busch, 2016). Proponents suggest that jurisdictional approaches create stronger incentive alignment for subnational governments and offer more durable pathways for financing and equity-sensitive benefit sharing, provided they are contextually grounded and designed with inclusive governance (Boyd et al., 2018; Fishbein & Lee, 2015).

J-REDD's most transformative potential lies in shifting the locus of mitigation from the commodification of abstract carbon units to the governance of land-based systems. This shift enables mitigation crediting to reflect territorial integrity, co-governance, and IPLC stewardship rather than narrow measurements of carbon flux (Sills et al., 2020). Yet J-REDD, like its predecessors, is not immune to failure.

For example, the East Kalimantan Jurisdictional Emissions Reduction Program (EK-JERP), the World Bank's flagship REDD+ pilot in Indonesia, institutionalized discrimination against Indigenous Peoples, particularly the Dayak Bahau community of Long Isun. EK-JERP's Benefit Sharing Plan imposed eligibility requirements that almost no Indigenous communities could meet, such as the need for formal legal recognition as a customary law community (Masyarakat Hukum Adat).

As of 2024, fewer than 1% of Indigenous Peoples in East Kalimantan had secured this status, effectively excluding them from REDD+ benefits (Forest Peoples Programme, 2024; Lang, 2025). Far from empowering Indigenous stakeholders, EK-JERP reinforced existing inequities by funneling finance through provincial structures while using technical eligibility rules to sideline customary landholders. These failings underscore how J-REDD – even at its most institutionally sophisticated – can replicate the very exclusion and credibility gaps it aims to resolve.

The effectiveness of J-REDD depends on consistent application of social and technical feasibility criteria and embedding climate action in territorially anchored systems of knowledge and consent (Forest Peoples Programme, 2024; Seymour & Busch, 2016; Sunderlin et al., 2015). Without this grounding, jurisdictional mitigation also risks becoming another iteration of the extractive logic it was meant to replace.

Reforming REDD+ Governance Is a Good First Step

Given the scale and operational experience of ongoing REDD+ projects, reforming REDD+ governance offers a pragmatic, though limited, entry point for improving outcomes. However, meaningful reform requires structural changes in how power, accountability, and incentives are organized; incremental adjustments will not suffice. Decentralizing REDD+ by empowering IPLCs to directly manage community-based programs could reduce dependency on national governments and external intermediaries. This shift demands genuine investment in securing land tenure, strengthening local land-use systems, and building IPLC capacity for co-design, monitoring, and financial management (UN-REDD Programme, 2021).

Pilot initiatives in Nepal and Mexico have shown that when IPLCs lead forest conservation, both social and ecological outcomes improve. In Nepal, community forest user groups have played a central role in managing forests and monitoring carbon stocks, while in Mexico, Indigenous communities have led REDD+ implementation aligned with territorial priorities (Angelsen et al., 2018; WWF, 2018). Still, these examples remain exceptions. Without reforms anchored in land tenure recognition, institutional transformation, and a deliberate transition away from reliance on intermediaries, REDD+ will continue to fall short. Strengthening IPLC autonomy must become a precondition – not a byproduct – of REDD+ if it is to move beyond its carbon-centric, extractive legacy (UN-REDD Programme, 2021).

The Great Green Wall: Cautionary Tale

Even large-scale, high-profile afforestation and reforestation programs raise significant concerns. China's Grain-for-Green initiative, one of the world's largest reforestation campaigns, has sequestered an estimated 2.5 $GtCO_2$ across more than 35 million hectares but has also triggered ecological simplification, the spread of invasive species, and widespread rural displacement – revealing the risks of scale-oriented, top-down approaches (Hua et al., 2016).

Similarly, Africa's GGW initiative, despite ambitious reforestation goals and international support, has struggled with ecological oversimplification, weak governance, and inconsistent community benefits (Raman, 2023). Both cases underscore that top-down, technocratic approaches to landscape restoration risk unintended social and ecological consequences unless local participation, adaptive management, and context-specific planning are prioritized. GGW is described in Box 3.2.

Box 3.2 The Great Green Wall – Ambition Without Anchoring

The Great Green Wall (GGW) initiative, launched to restore 100 million hectares of degraded land across the Sahel and improve rural livelihoods, has become a cautionary tale of how mismatched financial ambition, technocratic design, and fragmented delivery systems can undermine high-profile Nature-based Solutions (NbS) and stalled progress (Earth.Org., 2023).

Pledges versus Progress

Despite pledges of over $12 billion by 2025, implementation has lagged. By 2024, only around 10% of the target area had been restored – a modest increase from 4% in 2020 (UNCCD, African Union, & Pan-African Agency of the Great Green Wall, 2020). Much of the pledged funding failed to reach the community level, stymied by fragmented donor priorities, weak national coordination, and opaque financing pathways.

Effectiveness and Sustainability

Efforts to improve tree survival rates have shown very limited results. While resilient species and revised planting techniques offer promise, environmental stressors and governance failures continue to limit impact (UNCCD, 2025; Zhu et al., 2025). The project's carbon sequestration potential remains limited at the current scale, and land restoration outcomes vary widely across countries (FAO, 2025). Some carbon credit projects are also being marketed for forests that are not genuinely at risk, raising further questions about the legitimacy and additionality of such offset initiatives (The Guardian, 2024).

Equity and Inclusion Challenges

Livelihood improvements have largely accrued to better-connected regions, while marginalized communities report minimal benefits. Participatory governance mechanisms remain weak, tenure insecurity persists, and benefit-sharing arrangements are often opaque or exclusionary (UNEP, 2025). Up to 60% of funds have been absorbed by administrative and intermediary costs, including market fees, schedules, and transactions costs, limiting effectiveness on the ground (ART-TREES, 2024).

Lessons for Climate Finance and NbS

The GGW exemplifies the risks of carbon-focused, top-down NbS initiatives that lack accountability to Indigenous Peoples and Local Communities (IPLCs) (Knowable Magazine, 2023). Even with major financial pledges, meaningful outcomes remain elusive without socially legitimate, technically feasible STAM strategies and governance structures rooted in IPLC leadership, land rights, and context-appropriate definitions of success.

Challenges in Implementing NbS

Despite their rhetorical promise, NbS (NbS) still lack robust, context-specific evidence of effectiveness, sustainability, and equity at scale. This evidentiary gap hampers the translation of global ambitions into grounded, actionable outcomes (Cheng et al., 2023; Hou-Jones et al., 2021). In practice, many NbS initiatives unfold in what have been described as "expanding sacrifice zones" – regions where IPLCs are displaced, marginalized, or subjected to restrictive conservation regimes, often masking elite consolidation of land under the banner of climate action (Cabello & Gilbertson, 2012; Huff & Naess, 2022; IUCN, ENACT Partnership et al., 2024). The core issue is not climate mitigation itself, but the paternalistic implementation of externally conceived solutions.

Programs justified as serving the global commons are frequently imposed on the commons of the Global South – territories governed by plural tenure systems, embedded cultural values, and livelihood needs that top-down interventions rarely accommodate (Ostrom, 2009). As a result, IPLCs are effectively subsidizing the global community, with their lands diverted, labor extracted, and livelihoods compromised in exchange for "co-benefits" that are often deferred or never materialize (Garcia O'Donnell, 2024). Best practice for stakeholder engagement (Pacific Institute, 2022) are not remotely followed.

NbS interventions such as A/R or ecosystem restoration may yield visible environmental benefits, but these gains are unevenly distributed. PES programs, in particular, reveal deep disparities in benefit-sharing and participation (Cheng et al., 2023; Pagiola et al., 2005). Designing NbS that reflect local values, ecological dynamics, and governance realities requires a methodological pluralism that bridges disciplinary and policy silos (Akimowicz et al., 2023). Without such integration, rural development risks reproducing fragmented interventions that miss the social dimensions of feasibility.

Digital innovations such as AI and satellite-based monitoring are increasingly promoted to optimize NbS delivery in the Global South (Purwar & Akhter, 2023). However, without meaningful IPLC participation, these tools risk reinforcing data extractivism and top-down control over landscape governance. Climate solutions that are just must elevate IPLC rights, land tenure, and community governance, not merely technology (Allen, 2022; Rights and Resources Initiative, 2024).

Toward an Appropriate NbS Paradigm

More than a decade ago, *Redeeming REDD* (Brown, 2013) argued that REDD+ could become both effective and equitable – but only if its strategies, tools, and methods were reoriented toward social feasibility and institutional appropriateness. The core insight was that REDD+ needed to be evaluated not just by its technical potential to reduce emissions, but also by its ability to fit into complex landscapes shaped by power, rights, and governance. That vision of a "redeemable" REDD+ was compelling – but it was never realized. Ten years later, REDD+ remains hampered by carbon-centric logics, elite capture, and a persistent inability to address land tenure, IPLC governance, or meaningful participation at scale (Brown, 2013).

Today, NbS faces a similar legitimacy crisis. Despite its growing dominance in the discourse on land-based mitigation, NbS risks following REDD+ into the same cul-de-sac of technocratic control and social dislocation. The proliferation of GIS platforms, MRV systems, and climate finance pipelines has not led to structurally just outcomes for IPLCs. Instead, these tools have often reinforced top-down authority, producing maps without consent and metrics without meaning (Gabrysch et al., 2022; Turnhout et al., 2017).

But as with REDD+, the problem lies not with the tools themselves, but with the governance architectures in which they are embedded. GIS, participatory mapping, and spatial analysis can be empowering when placed in the hands of capacitated communities. Participatory geospatial platforms have enabled IPLCs to assert territorial claims, monitor deforestation, and engage governments on their own terms (Ballesteros-Bolaños et al., 2024; Sovacool et al., 2022). These are not merely technical upgrades; they are political acts that reposition IPLCs as knowledge holders and rights-bearers.

What is needed now is not the rejection of NbS, but the emergence of an appropriate NbS paradigm – one capable of receiving broad initial social license and designed around co-governance, legitimacy, and negotiated partnership. This paradigm would begin by grounding all interventions in genuine FPIC, not retroactive consultation or procedural box-checking. It would enable the creation of territorial negotiation platforms that bring together IPLCs, governments at multiple levels, the private sector, and aligned philanthropic and impact investors. These platforms would be accountable to IPLC priorities and embedded in legal and cultural frameworks, not merely project timelines.

As discussed in earlier chapters, an institution such as the Tropical Forest Forever Facility (TFFF) could potentially offer the infrastructure for this kind of partnership. While still hypothetical, the TFFF could serve as a braided space where finance, data systems, governance tools, and ecological knowledge converge to support long-term territorial governance led by IPLCs.

Under this model, the STAM framework would be reconfigured:

- **Strategies** would begin with IPLC visions of ecological care and territorial security, not carbon yield alone.
- **Tools** like GIS and remote sensing would support community decision-making, not substitute for it. For example, participatory geospatial platforms that allow

IPLCs to assert territorial claims, monitor deforestation, and engage governments, particularly through culturally adapted ethnoecological mapping tools in regions such as the Amazon Basin, should be promoted (Ballesteros-Bolaños et al., 2024).
- **Approaches** would prioritize inclusive co-governance over conditionality and donor-driven logic.
- **Methods** would be iterative and adaptive, co-designed through long-term processes that reflect cultural and institutional realities.

This is the paradigm that was imagined for REDD+ but never fully built. NbS now stands at a similar crossroads. Whether it, too, becomes redeemable will depend not on new technologies or financial instruments, but on the willingness of global actors to cede control, share power, and build climate solutions from the ground up.

Conclusion

NbS have become the dominant framework for land-based climate mitigation in the Global South, widely promoted as win-win interventions capable of delivering carbon, biodiversity, and development benefits (Gabrysch et al., 2022). Yet as this chapter has shown, the actual record across REDD+, NCS, AFOLU, A/R, PES, and even J-REDD reveals a consistent pattern: technocratic tools, carbon-centric metrics, and exclusionary governance structures routinely override the social, cultural, and political realities of the landscapes they claim to improve (Chausson et al., 2020; Cheng et al., 2023). Returning to the strong governance roots of community-based natural resource management – and taking them further – will be key to any future NbS success.

These are not isolated implementation failures but structural outcomes of a paradigm that privileges scale, quantification, and market efficiency over legitimacy, consent, and local feasibility (Gabrysch et al., 2022; Sheffield et al., 2022). Strategies continue to prioritize emissions reductions over tenure rights. Tools such as carbon accounting, GIS, and MRV dominate, while participatory governance remains an afterthought. Approaches are still top-down and donor-driven, and methods often extract data and compliance from IPLCs while offering little in return (Chausson et al., 2020; Cheng et al., 2023).

As a result, NbS interventions are too often imposed rather than co-created. Their benefits, when realized, disproportionately accrue to international actors, funders, and technical intermediaries, while IPLCs bear the costs: restricted access to land, disrupted livelihoods, and deferred or hypothetical co-benefits that frequently never arrive (Beaury et al., 2024b; Cheng et al., 2023).

A paradigm shift is not only desirable but also essential (Gabrysch et al., 2022; Sheffield et al., 2022). An appropriate NbS paradigm must reconfigure the STAM framework to foreground social and technical feasibility as the baseline for legitimacy:

- *Strategies* must begin with IPLC visions of stewardship and ecological care, not carbon yield.
- *Tools* must support – not substitute for – community decision-making and territorial autonomy.

- *Approaches* must prioritize inclusive co-governance over conditionality and donor logic and deliver measurable results.
- *Methods* must emerge from iterative, adaptive processes rooted in cultural values, tenure systems, and institutional complexity (Chausson et al., 2020; Gabrysch et al., 2022).

In most rural contexts of the Global South, sustainable climate outcomes will require achieving a mix of tangible benefits, since economic decision-making transcends simplistic profit maximization and climate "co-benefits" are just one of many challenges faced at the local level (Sheffield et al., 2022). Box 3.3 presents some of the more tangible challenges people face at local levels.

While NbS are widely promoted as delivering multiple "co-benefits," the empirical evidence supporting their contributions to non-carbon societal goals remains limited. Box 3.3 summarizes key societal challenges often associated with NbS, highlighting both theoretical claims and actual evidence gaps.

Despite claims of broad societal benefits, most NbS initiatives remain centered on carbon metrics and market-based valuations. Broader outcomes – such as equity, empowerment, and cultural integrity – are rarely built into monitoring frameworks, and where benefits are claimed, they are often anecdotal and not systematically evaluated (Beaury et al., 2024a; Chausson et al., 2020; Cheng et al., 2023). Future research and program design should move beyond carbon metrics to develop standardized but adaptable indicators for these underrepresented challenges, grounded in community-defined priorities and local governance systems (Gabrysch et al., 2022).

This shift is becoming visible in jurisdictional pilots, regenerative agriculture programs, and community-driven mapping initiatives but remains fragile and

Box 3.3 Underexplored Societal Dimensions of NbS

Societal Challenge	*Claimed NbS Contribution*	*Evidence Gap/Issue*
Food Security	Agroforestry and sustainable agriculture improve yields	Evidence is limited or context-specific; some A/R projects reduce food access (Mcgreevy et al., 2022).
Water Security	Forests and wetlands regulate water flow and quality	Hydrological outcomes are often not monitored; local trade-offs are underexplored.
Public Health	Ecosystem services reduce disease and improve nutrition	Few studies link NbS to measurable health outcomes.
Economic Development	Jobs and income from eco-tourism, PES, or restoration	Benefits often accrue to elites or intermediaries; marginal gains for IPLCs.
Climate Adaptation	Ecosystem resilience buffers against climate extremes	Evidence growing but often indirect; community perceptions underreported.

underfunded (Gabrysch et al., 2022). As later chapters will argue, the long-term credibility and effectiveness of NbS will hinge on deep reforms in finance, legal architecture, and data governance (Sheffield et al., 2022). Nature-based Solutions can only be successful if they prioritize IPLC leadership, secure land tenure, and shift from carbon metrics to a holistic STAM approach.

The aforementioned, controversial, aspirational platform that many hope could support such reform, the TFFF, is a proposed \$125 billion initiative led by Brazil to support the long-term preservation of tropical forests (Ministry of the Environment and Climate Change, 2025). Set to launch at COP30 in Belém in 2025, the TFFF aims to consolidate fragmented climate finance. However, its legitimacy will depend on whether it centers IPLC authority, enables negotiated access to finance, and is grounded in frameworks that reflect both social and technical feasibility across all STAM elements (TFFF, 2025). Later chapters will revisit this question, exploring why such a shift is not only just but also the most pragmatic path toward durable, effective land-based mitigation in the Global South.

NbS can indeed be a transformative tool for climate mitigation, but only if they evolve beyond the carbon-first frameworks that currently dominate. This requires rethinking their operational assumptions and methods. True transformation depends on genuine participation and co-governance, where IPLCs are empowered as decision-makers, not just participants. The future of NbS hinges on redefining success, not by carbon volumes or hectares conserved, but by how well they meet community needs and respect local governance systems to enable sustainable land use management systems to emerge.

References

Akimowicz, M., Robinson, C., & Fortin, M. J. (2023). Ensuring inclusion in nature-based solutions: Strategies for fair benefit-sharing. *Sustainability Science*, 18(1), 93–107.

Alatorre, G., & Frenk, G. (2000). The Mexican ejido: Lessons for rural development policy and natural resource management. *World Development*, 28(5), 761–782.

Allen, K. (2022). Indigenous Peoples and communities drive climate finance reform. *Mongabay*. https://news.mongabay.com/2022/12/indigenous-peoples-and-communities-drive-climate-finance-reform/

Angelsen, A., Martius, C., De Sy, V., Duchelle, A. E., Larson, A. M., & Pham, T. T. (2018). *Transforming REDD+: Lessons and New Directions*. CIFOR.

ART-TREES. (2024). Fee schedule. https://www.artredd.org/resources.

Badgley, G., Freeman, J., Hamman, J. J., Haya, B., & Cullenward, D. (2022). *California's Forest Carbon Offsets Buffer Pool Is Severely Undercapitalized*. CarbonPlan.

Ballesteros-Bolaños, H., et al. (2024). Ethnoecological participatory mapping tools for Indigenous land governance in the Amazon basin. *Sustainability*, 16(19), 8524.

Beaury, E. M., Smith, J., & Levine, J. M. (2024a). Global suitability and spatial overlap of land-based climate mitigation strategies. *Global Change Biology*. https://pubmed.ncbi.nlm.nih.gov/39319461/

Beaury, E., Streck, C., Ajani, J., Benton, T., Duchelle, A., Funk, J., . . . & Zarin, D. (2024b). The conflicting practices in using land to tackle climate change. *Carbon Brief*. https://www.carbonbrief.org/guest-post-the-conflicting-practices-in-using-land-to-tackle-climate-change/

Bee, B.A. (2019). Recentralising Political Power Through Decentralised Environmental Governance: A Case from Mexico's Early REDD+ Program. *Conservation & Society*, 17(1), 96–107. https://doi.org/10.4103/cs.cs_17_111

Berkes, F. (2008). *Sacred Ecology: Traditional Ecological Knowledge and Resource Management* (2nd ed.). Routledge.

BioCarbon Fund ISFL. (2024). *Final Synthesis Report: BioCarbon Fund Initiative for Sustainable Forest Landscapes (ISFL) Independent Evaluation*. World Bank. https://www.biocarbonfund-isfl.org/sites/default/files/2024-06/2.%20ISFL%20BioCarbonEvaluation_Final%20Synthesis%20Report_April%2030%2C%202024_Final.pdf

Boyd, W., Stickler, C. M., Duchelle, A. E., Seymour, F., Nepstad, D., Bahar, N. H. A., & Rodriguez-Ward, D. (2018). Jurisdictional approaches to REDD+ and low emissions development: Progress and prospects. *Working Paper*. Washington, DC: World Resources Institute. https://gcftf.org/wp-content/uploads/2020/12/ending-tropical-deforestation-jurisdictional-approaches-redd.pdf

Bray, D. B., Antinori, C., & Torres-Rojo, J. M. (2008). The Mexican model of community forest management: The role of agrarian policy, forest policy and entrepreneurial organization. *Forest Policy and Economics, 10*(6), 407–417.

Brown, M. (2013). *Redeeming REDD: Policies, Incentives and Social Feasibility for Avoided Deforestation*. Routledge. https://doi.org/10.4324/9780203123652

Cabello, J., & Gilbertson, T. (2012). A colonial mechanism to enclose lands: A critical review of two REDD+-focused special issues. *Ephemera: Theory & Politics in Organization*, 12(2), 162–180.

Carbon Market Watch. (2023). *The Great Carbon Market Divide: Why Offset Projects Fail to Deliver for Communities*. Carbon Market Watch.

Chausson, A., Turner, B., Seddon, D., Chabaneix, N., Girardin, C. A. J., Kapos, V., Key, I., Roe, D., Smith, A., Woroniecki, S., & Seddon, N. (2020). Mapping the effectiveness of nature-based solutions for climate change adaptation. *Global Change Biology*, 26(11), 6134–6155. https://doi.org/10.1111/gcb.15310

Cheng, S. H., Costedoat, S., Sigouin, A., Calistro, G. F., Chamberlain, C. J., Lichtenthal, P., . . . & Brenes, C. L. M. (2023). Assessing evidence on the impacts of nature-based interventions for climate change mitigation: A systematic map of primary and secondary research from subtropical and tropical terrestrial regions. *Environmental Evidence*, 12(1). https://doi.org/10.1186/s13750-023-00312-3

CIFOR-ICRAF. (2024). *In the Spotlight: Presenting CIFOR-ICRAF's Ongoing Commitment to Resilient Tree, Forest, and Agroforestry Landscapes*. Annual Report 2024.

Cohen-Shacham, E., Walters, G., Janzen, C., & Maginnis, S. (Eds.). (2016). *Nature-Based Solutions to Address Global Societal Challenges*. IUCN, Gland, Switzerland. https://doi.org/10.2305/IUCN.CH.2016.13.en

Collins, N., B. van Zanten, I. Onah, L. Marsters, L. Jungman, R. Hunter, N. von Turkovich, J. Anderson, G. Vidad, T. Gartner, & B. Jongman. (2025). Growing Resilience: Unlocking the Potential of Nature-Based Solutions for Climate Resilience in Sub-Saharan Africa. © World Bank.

Dawson, N. M., Coolsaet, B., Sterling, E. J., Loveridge, R., Gross-Camp, N. D., Wongbusarakum, S., . . . & Rosado-May, F. J. (2021). The role of Indigenous Peoples and local communities in effective and equitable conservation. *Ecology and Society*, 26(3). https://doi.org/10.5751/ES-12625-260320

da Silva, R.F.B., de Castro Victoria, D., Nossack, F.Á. et al. (2023). Slow-down of deforestation following a Brazilian forest policy was less effective on private lands than in all conservation areas. *Commun Earth Environ.*, 4, 111. https://doi.org/10.1038/s43247-023-00783-9

D.K., & Kamukama, M. (2022). Trees for Global Benefit Uganda: A case study on the failures of carbon offsetting. Companies purchasing carbon credits as offsets are guilty of greenwashing and inflicting adverse impacts on local communities. *Global Forest Coalition*. https://globalforestcoalition.org/wp-content/uploads/2022/10/Informe_Uganda_2-1.pdf

Dooley, K., Keith, H., Larson, A.M., Catacora-Vargas, G., Carton, W., Christiansen, K.L., Enokenwa, O., Frechette, A., Hugh, S., Ivetic, N., Lim, L.C., Lund, J.F., Luqman, M., Mackey, B., Monterroso, I., Ojha, H., Perfecto, I., Riamit, K., Robiou du Pont, Y., & Young, V. (2022a). *The Land Gap Report 2022*. LandGap.org; Center for International Forestry Research (CIFOR-ICRAF).

Dooley, K., Streck, C., Tsioumani, E., & Sharma, S. (2022b). Rights, carbon, and the politics of nature-based solutions. *Global Sustainability*, 5(e3), 1–14.
Duchelle, A. E., Simonet, G., Sunderlin, W. D., & Wunder, S. (2018). What is REDD+ achieving on the ground? Current status and lessons for the future. *Current Opinion in Environmental Sustainability*. https://www.sciencedirect.com/science/article/pii/S1877343517301872
Dunlop, T., Khojasteh, D., Cohen-Shacham, E., et al. (2024). The evolution and future of research on nature-based solutions to address societal challenges. *Communications Earth & Environment*, 5, 132. https://doi.org/10.1038/s43247-024-01308-8
Earth.Org. (2023). The Great Green Wall Is Failing, But its Legacy Could Still Be A Success. *Earth.Org*. https://earth.org/the-great-green-wall-legacy/
Ecosystem Marketplace. (2021). State of the Voluntary Carbon Markets 2021: Installment 1. Forest Trends. Retrieved August 20, 2025, from https://www.ecosystemmarketplace.com/publications/state-of-the-voluntary-carbon-markets-2021/
Ecosystem Marketplace. (2024). 2024 State of the Voluntary Carbon Market (SOVCM). https://www.ecosystemmarketplace.com/publications/2024-state-of-the-voluntary-carbon-markets-sovcm/
EIT Food. (2023). *Annual reporting 2023*. Retrieved August 20, 2025, from https://reporting.eitfood.eu/annual-reporting-data-2024
European Commission: Directorate-General for Research and Innovation. (2021). Evaluating the impact of nature-based solutions – A summary for policy makers. https://www.naturebasedsolutionsinitiative.org/news/european-commission-report-nature-based-solutions/
FAO. (2021). The state of the world's land and water resources for food and agriculture: Systems at breaking point. https://www.fao.org/documents/card/en/c/cb7654en
Fishbein, G., & Lee, D. (2015). *Early Lessons from Jurisdictional REDD+ and Low Emissions Development Programs*. Arlington, VA: The Nature Conservancy. Retrieved August 20, 2025, from https://www.nature.org/media/climatechange/REDD+_LED_Programs.pdf
Food and Agriculture Organization (FAO). (2025). *Monitoring Land Degradation and Restoration in Africa's Drylands*. FAO.
Friess, D. A., Thompson, B. S., Brown, B., Amir, A. A., Cameron, C., & Sidik, F. (2016). Policy challenges and approaches for the conservation of mangrove forests in Southeast Asia. *Conservation Biology*, 30(5), 933–949. https://doi.org/10.1111/cobi.12784
Forest & Climate Leaders' Partnership Platform. (2024). PLATFORM TO SUPPORT INDIGENOUS PEOPLES AND LOCAL COMMUNITIES IN CLIMATE AND FOREST ACTION: Work Plan 2024–2025. Retrieved from https://forestclimateleaders.org/wp-content/uploads/2025/01/IPLC-Platform-Work-Plan-24_09_09.pdf
Forest Peoples Programme. (2023). *Wrong Turn in Peru? How Forest Carbon Deals Threaten Indigenous Territories*. Forest Peoples Programme Report.
Forest Peoples Programme, Kawari, Perkumpulan Nurani Perempuan, & Re:wild. (2024, November 12). *REDD+ Alert: How the World Bank's Jurisdictional REDD+ Pilot in East Kalimantan Discriminates Against Indigenous Peoples*. Forest Peoples Programme.
Friess, D. A., Thompson, B. S., Brown, B., Amir, A. A., Cameron, C., Koldewey, H. J., Sasmito, S. D., Sidik, F., & Worthington, T. A. (2016). Policy challenges and approaches for the conservation of mangrove forests in Southeast Asia. *Conservation Biology*, 30(5), 933–949.
Gabrysch, S., Jadin, J. L., & Chan, K. M. A. (2022). Integrating institutional and governance dimensions into spatial planning tools for climate mitigation. *Science of the Total Environment*, 846, 157304.
Garcia O'Donnell, A. (2024). Green sacrifice zones: The justice implications of the green transition. *TASC*. https://www.tasc.ie/blog/2024/02/06/green-sacrifice-zones-the-justice-implications-of/
Garnett, S. T., Burgess, N. D., Fa, J. E., Fernández-Llamazares, Á., Molnár, Z., Robinson, C. J., . . . & Watson, J. E. M. (2018). A spatial overview of the global importance of Indigenous lands for conservation. *Nature Sustainability*, 1(7), 369–374. https://www.nature.com/articles/s41893-018-0100-6

General Mills. (2023). Regenerative agriculture. https://www.generalmills.com/how-we-make-it/healthier-planet/environmental-impact/regenerative-agriculture

Girardin, C. A. J., Jenkins, S., Seddon, N., Allen, M., Lewis, S., Wheeler, C. E., . . . & Griscom, B. (2021). Nature-based solutions can help cool the planet – if we act now. *Nature*, 593(7858), 191–194. https://doi.org/10.1038/d41586-021-01241-2

Global Witness. (2024). *The Carbon Con: Indigenous Land, Carbon Offsets, and Pressure Tactics in Latin America and Asia*. Global Witness Special Report.

Green Climate Fund. (2024). *Annual Portfolio Performance Report 2023*. https://www.greenclimate.fund/publications/portfolio-performance-report-2023

Gvein, M., Bager, S. L., Mauser, W., & Rabinovich, J. (2023). Navigating the contradictions of nature-based solutions: Competing discourses and sustainability outcomes. *Environmental Science & Policy*, 145, 38–47.

Hou-Jones, X., Roe, D., & Holland, E. (2021). *Nature-based Solutions in Action: Lessons from the Frontline*. London. Bond.

Hua, F., Wang, X., Zheng, X., Fisher, B., Wang, L., Zhu, J., ... & Wilcove, D. S. (2016). Opportunities for biodiversity gains under the world's largest reforestation programme. *Nature Communications, 7,* 12717. https://doi.org/10.1038/ncomms12717

Huff, A., & Naess, L. O. (Eds.). (2022). Introduction: Reframing climate and environmental justice. *IDS Bulletin*, 53(4).

Intergovernmental Panel on Climate Change (IPCC). (2022). Climate change 2022: Mitigation of climate change. Contribution of Working Group III to the Sixth Assessment Report of the Intergovernmental Panel on Climate Change. Cambridge University Press. https://www.ipcc.ch/report/ar6/wg3/

ISEAL Alliance. (2020). Getting FPIC Right: How voluntary standards can go further. https://www.isealalliance.org/sustainability-news/getting-fpic-right-how-voluntary-standards-can-go-further

IUCN. (2020). *Global Standard for Nature-Based Solutions: A User-Friendly Framework for the Verification, Design and Scaling up of NbS*. IUCN. https://doi.org/10.2305/IUCN.CH.2020.08.en

IUCN. (2022). Nature-based solutions for climate. *International Union for Conservation of Nature*. https://www.iucn.org/our-work/topic/nature-based-solutions-climate

IUCN, ENACT Partnership, et al. (2024). State of ENACT NbS Goals Report: Year One Roadmap. *International Union for Conservation of Nature (IUCN) & ENACT Partnership (including TASC and other stakeholders)*. https://iucn.org/sites/default/files/2024-02/iucn_state-of-enact-nbs-goals-report.pdf

Kabisch, N., Korn, H., Stadler, J., & Bonn, A. (2022). *Nature-Based Solutions to Climate Change Adaptation in Urban Areas: Linkages between Science, Policy and Practice*. Springer Nature. https://link.springer.com/book/10.1007/978-3-319-56091-5

Kaiser, J., Haase, D., & Krueger, T. (2023). Collective payments for ecosystem services: a counterpart of commodification and privatization trends in nature conservation? *Ecology and Society*, 28(1), Article 13. https://ecologyandsociety.org/vol28/iss1/art13/

Kengoum, F., Pham, T. T., & Mihigo, B.-P. N. (2022). From participation to inclusive forest governance in REDD+ in the DRC (InfoBrief No. 363). CIFOR-ICRAF. https://doi.org/10.17528/cifor/008498

Kill, J. (2016). The Kasigau Corridor REDD+ Project in Kenya: A crash dive for Althelia Climate Fund. *Counter Balance & Re:Common, in cooperation with Jamaa Resource Initiative*. https://counter-balance.org/uploads/files/Reports/Flagship-Reports-Files/2017-The-Kasigau-Corridor-REDD-Kenya.pdf

Kimeu, C. (2025). Netflix and Meta's carbon credits snared in dispute with Maasai herders. *The Wall Street Journal*. https://www.wsj.com/us-news/climate-environment/netflix-and-metas-carbon-credits-snared-in-dispute-with-maasai-herders-4f01a59f?msockid=2bb2e73fd6a56cbf28acf2f3d7416d2e

Knowable Magazine. (2023). Abandon the idea of 'great green walls'. *Knowable Magazine*. https://knowablemagazine.org/content/article/society/2023/abandon-idea-great-green-walls

Lang, C. (2025, March 27). REDD+ Alert: World Bank discrimination against Indigenous Peoples in the East Kalimantan jurisdictional REDD project. *REDD-Monitor*. https://reddmonitor.substack.com/p/redd-alert-world-bank-discrimination

Le, T. T., Vodden, K., Wu, J., Bullock, R., & Sabau, G. (2023). Payments for ecosystem services programs: A global review of contributions towards sustainability. *Heliyon, 10*(1), e22361. https://doi.org/10.1016/j.heliyon.2023.e22361

Lohmann, L. (2009). Regulation as corruption in the carbon offset markets. In S. Böhm & S. Dabhi (Eds.), *Upsetting the Offset: The Political Economy of Carbon Markets* (pp. 175–192). MayFlyBooks.

López, J. E., & Shapiro, D. (2018). Agroforestry and sustainable livelihoods: Balancing agricultural productivity and ecosystem services. In F. Montagnini (Ed.), *Integrating Landscapes: Agroforestry for Biodiversity and Ecosystem Services* (pp. 51–65). Springer.

Mace, G. M. Whose conservation? *Science*, 345, 1558–1560 (2014).

Mackenzie, C. (2016). Comparing voluntary carbon market project design: Standards, entrepreneurs, and the marketization of nature. *Environment and Planning A*, 48(12), 2345–2363.

Marshall, C. (2023, November 9). Kenya's Ogiek people being evicted for carbon credits – lawyers. *BBC News*. https://www.bbc.com/news/world-africa-67352067

Marten, G. G., Panday, P., Coe, M. T., & Lathuillière, M. J. (2019). Land use changes in the Southeastern Amazon and trends in rainfall and river flow: Xingu River Basin. *Earth's Future*, 7(8), 919–935.

Massarella, K., Sallu, S. M., Ensor, J. E., & Marchant, R. (2018). REDD+, hype, hope and disappointment: The dynamics of expectations in conservation and development pilot projects. *World Development*, 109, 375–385. https://doi.org/10.1016/j.worlddev.2018.05.006

Mcgreevy, S. R., et al. (2022). Sustainable agrifood systems for a post-growth world. *Nat. Sustain*, 5, 1011–1017.

McKinsey & Company. (2022). *A Blueprint for Scaling Voluntary Carbon Markets to Meet the Climate Challenge*. Sustainability Practice, McKinsey & Company.

Milne, M. (2005). Transaction costs of forest carbon projects. Working Paper CC05. Report submitted to the University of New England as part of the ACIAR Project ASEM/1999/093, The role of carbon sequestration credits in influencing the economic performance of farm forestry systems. Center for International Forestry Research (CIFOR), Bogor, Indonesia. Retrieved from https://www.ecosystemmarketplace.com/wp-content/uploads/archive/documents/Doc_355.pdf and http://www.une.edu.au/febl/Econ/carbon/

Ministry of the Environment and Climate Change (Brazil). (2025, July 1). Tropical Forests Forever Fund boosts international support and positions itself as a new global model for climate finance. *Forests: COP30 News*.

Mukpo, A. (2025, May 15). Kenyan soil carbon project suspended for a second time. *Mongabay*. https://news.mongabay.com/short-article/kenyan-soil-carbon-project-suspended-for-a-second-time/

Muradian, R., Corbera, E., Pascual, U., Kosoy, N., & May, P. H. (2010). Reconciling theory and practice: An alternative conceptual framework for understanding payments for environmental services. *Ecological Economics,* 69(6), 1202–1208. https://doi.org/10.1016/j.ecolecon.2009.11.006

Ostrom, E. (1990). *Governing the Commons: The Evolution of Institutions for Collective Action*. Cambridge University Press.

Ostrom, E. (2009). *A General Framework for Analyzing Sustainability of Social-Ecological Systems*. Science, 325(5939), 419–422. https://doi.org/10.1126/science.1172133

Pacific Institute. (2022). Stakeholder engagement guide for nature-based solutions. *Pacific Institute*. https://pacinst.org/stakeholder-engagement-guide-for-nature-based-solutions/

Pagiola, S., Arcenas, A., & Platais, G. (2005). Can Payments for Environmental Services Help Reduce Poverty? An Exploration of the Issues and the Evidence to Date from Latin America. *World Development, 33*(2), 237–253. https://doi.org/10.1016/j.worlddev.2004.07.011

Papandreou, A. (2016). Markets, morals and the future of nature-based solutions. *Ecological Economics*, 131, 218–227.

Porras, I., Grieg-Gran, M., & Neves, N. (2013a). Watershed services for agriculture: A review of evidence on the linkages between natural resource management and the performance of agricultural production systems. *IIED*. https://pubs.iied.org/16559iied

Porras, I., Neves, N., & Barlow, J. (2013b). *The Role of Payment for Ecosystem Services (PES) in Promoting Sustainable Land Use: Evidence from Latin America and Sub-Saharan Africa*. Earthscan.

Probst, B.S., Toetzke, M., Kontoleon, A. et al. Systematic assessment of the achieved emission reductions of carbon crediting projects. *Nat Commun 15*, 9562 (2024). https://doi.org/10.1038/s41467-024-53645-z

Purwar, A., & Akhter, Z. (2023). Enabling nature-based solutions with AI. *Blue Sky Analytics*. https://blueskyhq.io/blog/enabling-nature-based-solutions-in-global-south-with-satellite-data-and-ai

Raman, S. (2023, August 3). Progress is slow on Africa's Great Green Wall, but some bright spots bloom. *Mongabay Environmental News*. https://news.mongabay.com/2023/08/progress-is-slow-on-africas-great-green-wall-but-some-bright-spots-bloom/

Reyes-García, V., et al. (2022). Nature-based solutions: Opportunities and challenges for Indigenous peoples and local communities. *Nature Sustainability*, 5, 869–876.

Reytar, K., & Veit, P. G. (2017). 5 maps show how important Indigenous Peoples and local communities are to the environment. *World Resources Institute*. https://www.wri.org/insights/5-maps-show-how-important-indigenous-peoples-and-local-communities-are-environment

Rights and Resources Initiative. (2015a). *The Land Tenure Dilemma in REDD+ and PES Mechanisms: Implications for Indigenous and Local Communities*. Rights and Resources Initiative.

Rights and Resources Initiative. (2015b). *Who Owns the World's Land? A Global Baseline of Formally Recognized Indigenous and Community Land Rights*. Washington, DC: RRI.

Rights and Resources Initiative. (2023). *Who Owns the World's Land?* (2nd ed.). https://rightsandresources.org

Rodale Institute. (2020). Regenerative Agriculture and the Soil Carbon Solution. Rodale Institute White Paper. Retrieved August 20, 2025, from https://rodaleinstitute.org/education/resources/regenerative-agriculture-and-the-soil-carbon-solution/

Roe, S., Streck, C., Beach, R., Busch, J., Chapman, M., Engelmann, J., Franks, M., & Funk, J. (2021). Land-based measures to mitigate climate change: Potential and feasibility by country. *Global Change Biology*, 27(23), 6025–6058. https://doi.org/10.1111/gcb.15873

Scheidel, A., & Work, C. (2018). Forest plantations and climate change discourses: New powers of "green" grabbing in Cambodia. *Land Use Policy*, 77, 9–18. https://doi.org/10.1016/j.landusepol.2018.04.057

Seddon, N., Chausson, A., Berry, P., Girardin, C. A. J., Smith, A., & Turner, B. (2020). Understanding the value and limits of nature-based solutions to climate change and other global challenges. *Philosophical Transactions of the Royal Society B*, 375(1794), 20190120. https://doi.org/10.1098/rstb.2019.0120

Seddon, N., Smith, A., Smith, P., Key, I., Chausson, A., Girardin, C., . . . & Turner, B. (2021). Getting the message right on nature-based solutions to climate change. *Global Change Biology*, 27(8), 1518–1546. https://doi.org/10.1111/gcb.15475

Seymour, F., & Busch, J. (2016). Why forests? Why now? The science, economics, and politics of tropical forests and climate change. *Center for Global Development*. https://www.cgdev.org/publication/why-forests-why-now

SGS. (2024). Sustainability solutions: Water conservation in agriculture. *SGS SA*. https://www.sgs.com/-/media/sgscorp/documents/corporate/brochures/sgs-susty-sustainability-solutions-en.cdn.en-BE.pdf

Sills, E. O., Atmadja, S., de Sassi, C., Duchelle, A. E., Kweka, D. L., Resosudarmo, I. A. P., & Sunderlin, W. D. (Eds.). (2020). *REDD+ on the Ground: A Case Book of Subnational*

Initiatives Across the Globe. CIFOR. https://www.cifor.org/publications/pdf_files/Books/BSills1501.pdf

Snilsveit, B., Stevenson, J., Langer, L., da Silva, N., Rabat, Z., et al. (2019). Incentives for climate mitigation in the land use sector: The effects of payment for environmental services (PES) on environmental and socio-economic outcomes in low- and middle-income countries. *3ie Systematic Review*, 44. https://doi.org/10.23846/SR00044

Snyder, K. (2023). Land-based mitigation strategies and their implications for local communities in the Global South. *Lincoln Institute of Land Policy Working Paper, WP23KS1*. https://www.lincolninst.edu/publications/working-papers/land-based-mitigation-strategies-implications-local-communities-global-south

Sobrevila, C. (2008). *The Role of Indigenous Peoples in Biodiversity Conservation: The Natural but Often Forgotten Partners*. Washington, DC: World Bank.

Sovacool, B. K., Brisbois, M. C., & Foley, R. (2022). The political economy of participatory mapping and climate justice. *Climate Policy*, 22(1), 12–31.

Sterling, E. J., Pascua, P. A., Sigouin, A., Gazit, N., Macey, S., Betley, E., . . . & McCarter, J. (2017). Biocultural approaches to well-being and sustainability indicators across scales. *Nature Ecology & Evolution*, 1(12), 1798–1806. https://doi.org/10.1038/s41559-017-0349-6

Sunderlin, W. D., Sills, E. O., Duchelle, A. E., Ekaputri, A. D., Kweka, D., Toniolo, M. A., Ball, S., Doggart, N., Pratama, C. D., Padilla, J. T., Enright, A., & Otsyina, R. M. (2015). REDD+ at a critical juncture: assessing the limits of polycentric governance for achieving climate change mitigation. *International Forestry Review, 17*(4), 400–413. https://doi.org/10.1505/146554815817476468

Survival International. (2023). *Blood Carbon: How a Carbon Offset Scheme Makes Millions from Indigenous Land in Northern Kenya*. Survival International.

Sustainable Markets Initiative Agribusiness Task Force. (2025). Scaling regenerative farming: An action plan. https://www.sustainable-markets.org/news/new-sustainable-markets-initiative-guide-addresses-the-challenges-to-scaling-regenerative-farming

Sutherland, L. (2021, November 2). $1.7 billion pledged in support of Indigenous and local communities' land tenure. *Mongabay*. https://news.mongabay.com/2021/11/1-7-billion-pledged-in-support-of-indigenous-and-local-communities-land-tenure/

Tauli-Corpuz, V., Alcorn, J., Molnar, A., Healy, C., & Barrow, E. (2020). Cornered by PAs: Adopting rights-based approaches to enable cost-effective conservation and climate action. *World Development*, 130, 104923. https://doi.org/10.1016/j.worlddev.2020.104923

The Guardian. (2023). Revealed: more than 90% of rainforest carbon offsets by biggest certifier are worthless, analysis shows. https://www.theguardian.com/environment/2023/jan/18/revealed-forest-carbon-offsets-biggest-provider-worthless-verra-aoe

The Guardian. (2024). BP-owned company is selling carbon credits on trees that aren't in danger, analysis finds. https://www.theguardian.com/environment/article/2024/jul/11/finite-carbon-forest-offsets-analysis

Turnhout, E., Neves, K., & de Lijster, E. (2017). "Measurementality" in biodiversity governance: Knowledge, transparency, and the intergovernmental science-policy platform on biodiversity and ecosystem services (IPBES). *Environment and Planning A*, 49(4), 645–664.

UNEP. (2022). Resolution Adopted by the United Nations Environment Assembly on 2 March 2022. 5/5. *Nature-Based Solutions for Supporting Sustainable Development*. United Nations Environment Assembly of the United Nations Environment Programme. UNEP, Nairobi.

United Nations Convention to Combat Desertification (UNCCD), African Union, & Pan-African Agency of the Great Green Wall. (2020). The Great Green Wall: Implementation status and way ahead to 2030 (with technical inputs from Climatekos gGmbH). UNCCD. https://catalogue.unccd.int/1551_GGW_Report_ENG_Final_040920.pdf

United Nations Convention to Combat Desertification (UNCCD). (2025). *Great Green Wall Implementation and Governance Review*. UNCCD.

United Nations Environment Programme (UNEP). (2025). *Equity and Access in Land Restoration Finance: Challenges in the Sahel Region*. UNEP.

UN-REDD Programme. (2021). UN-REDD Programme: Annual Report 2021. *United Nations Environment Programme (UNEP), United Nations Development Programme (UNDP), Food and Agriculture Organization of the United Nations (FAO)*. Retrieved August 20, 2025, from https://www.un-redd.org/our-work/projects

Veit, P. G., & Reytar, K. (2017). By the numbers: Indigenous and community land rights. *World Resources Institute*. https://www.wri.org/insights/numbers-indigenous-and-community-land-rights

Verra. (2025). VCU issuance levy fee schedule. https://verra.org

Voluntary Carbon Markets Integrity Initiative (VCMI). (2023). Beyond Beneficiaries: Fairer Carbon Market Frameworks. https://vcmintegrity.org/wp-content/uploads/2023/08/TNC_Beyond-Beneficiaries-030823.pdf

Walker, W. S., Gorelik, S. R., Baccini, A., Aragon-Osejo, J. L., Josse, C., Meyer, C., . . . & Schwartzman, S. (2020). The role of forest conversion, degradation, and disturbance in the carbon dynamics of Amazon Indigenous territories and protected areas. *Proceedings of the National Academy of Sciences*, 117(6), 3015–3025. https://www.pnas.org/doi/full/10.1073/pnas.1913321117

Waweru, E. (2016, June 22). Some remarks on environmental technologies from REDD+ exchange 2016. *Global Policy Journal*. https://www.globalpolicyjournal.com/blog/22/06/2016/some-remarks-environmental-technologies-redd-exchange-2016

West, T. A. P., et al. (2023). Action needed to make carbon offsets from forest conservation work for climate change mitigation. *Science*, 381(6657), 633–636.

Wiggins, S., Randriamalala, J. R., & Ratsimbazafy, J. (2018). Bushmeat hunting and use in the Makira Forest, North-Eastern Madagascar: A conservation and livelihoods issue. *Oryx*, 52(2), 321–331.

Wildlife Works. (2024). Kasigau Corridor REDD+ Project, Kenya. Retrieved May 9, 2025, from https://www.wildlifeworks.com/redd-projects/kasigau-kenya.

World Bank. (2022). Strengthening Indigenous land tenure: A critical pillar of climate action. *Forest Carbon Partnership Facility*. https://www.forestcarbonpartnership.org/news-story/strengthening-indigenous-land-tenure-critical-pillar-climate-action

World Resources Institute. (2024). Climate cooperation disparities: An analysis of the Global South's limited participation in intergovernmental cooperation on climate change mitigation. https://www.wri.org/research/climate-cooperation-disparities-analysis-global-south-climate-change-mitigation

Wunder, S. (2005). *Payments for Environmental Services: Some Nuts and Bolts*. CIFOR. https://www.cifor.org/publications/pdf_files/OccPapers/OP-42.pdf

WWF. (2013). Indigenous Peoples, local communities and REDD+. *World Wildlife Fund*. https://wwf.panda.org/wwf_news/?202295/Indigenous-Peoples-Local-Communities-and-REDD

WWF. (2018). Living Planet Report 2018: Aiming Higher. *World Wide Fund for Nature (WWF)*, Gland, Switzerland. https://www.wwf.org.uk/sites/default/files/2018-10/LPR2018_Full%20Report.pdf

Zhu, A. L., Ndiaye, A., Dahm, R., Mauclaire, M., & Boas, I. (2025). Africa's great green mirage? Assessing the disconnect between global finance and local implementation in Africa's Great Green Wall. *Land Use Policy, 127,* 107670. https://doi.org/10.1016/j.landusepol.2025.107670

4 The Foundations of Failure – Tenure Insecurity, Governance Breakdown, and the Crisis of Climate Legitimacy

Nature-based Solutions (NbS) have become the flagship strategies for land-based climate mitigation across the Global South. Yet, despite widespread commitments to justice, inclusion, and sustainability, these efforts often fall short during implementation. This chapter examines three interlinked barriers that hinder the effective execution of NbS: insecure land tenure, weak institutional capacity, and administrative corruption (Stevens et al., 2017). In particular, securing land tenure is recognized as essential to advance equitable land rights, combating corruption, and enable achieving the Sustainable Development Goals (Transparency International, 2023). It also is key to biodiversity conservation, as IPLCs steward 25% of global lands and 80% of forest biodiversity (Forest Peoples Programme et al., 2020).

These are not mere technical obstacles but structural failures embedded within governance systems that frequently overlook IPLCs, marginalizing their rights (Larson et al., 2022; Larson & Soto, 2008; Lind et al., 2020; Reyes-García et al., 2022; Rights and Resources Initiative, 2021; Tauli-Corpuz et al., 2020). The following sections will explore how each of these barriers – land tenure insecurity, weak governance, and corruption – manifests in real-world contexts, using global evidence (Rights and Resources Initiative, 2022; Scoones, 1995) and regional case studies to highlight their impact on the effectiveness of land-based climate mitigation, with Muroa (2022) offering an excellent example from Papua New Guinea.

As explored in Chapters 1 and 3, these foundational issues – ranging from governance breakdowns to flawed ESG frameworks – are central to the critique of current climate mitigation paradigms and their failure to achieve equitable outcomes.

The aforementioned barriers are not just technical but deeply structural, undermining the very foundation upon which NbS rest. Unless these structural barriers are confronted, the promises of climate justice, equity, and sustainability in land-based mitigation will remain elusive, undermining both the legitimacy and feasibility of global climate action (Blackman et al., 2017; Garcia O'Donnell, 2024; IPES-Food, 2024).

These communities often govern land through reciprocal obligations, kinship-based access, and collective tenure arrangements that do not map easily onto formalized, market-oriented systems of climate finance and land commodification (Adhikari & Boon, 2010; Agrawal & Gibson, 1999). Even research from outside the Global South, such as Adhikari & Boon's study of the Southern

DOI: 10.4324/9781003380320-5

Appalachians, underscores how ignoring these embedded social systems can derail conservation goals. Similar dynamics are even more pronounced in the Global South, where customary governance is central to land-based climate mitigation.

What is most striking, as numerous authors have noted in recent years, is that tenure-secure forests under Indigenous and community stewardship sequester two to three times more carbon than state-managed or unprotected areas, storing an estimated 293 billion metric tons of carbon – the equivalent of 33 years of global energy emissions (Gray et al., 2015; Gumucio, 2024; Nepstad et al., 2019; Project Drawdown, 2024; Rights and Resources Initiative, 2018; Robinson et al., 2014). The feasibility and effectiveness of land-based climate solutions are thus directly tied to the security of land tenure (Roe et al., 2021). The absence of clear ownership or use rights creates uncertainty and disincentivizes long-term stewardship, which is critical for initiatives like REDD+ that depend on sustained land management practices (Massarella et al., 2018; Masuda et al., 2020; Rights and Resources Initiative, 2024; Sunderlin et al., 2018; Tseng et al., 2021; Walker et al., 2024).

As outlined in Chapters 1 and 3, such systemic distortions erode the legitimacy, feasibility, and equity of prevailing climate mitigation paradigms, particularly those rooted in technocratic models that exclude local perspectives from decision-making. They also undermine IPLC biodiversity conservation contributions (Forest Peoples Programme, et al., 2020). Unless addressed directly, these breakdowns will continue to delegitimize climate and conservation action, and reproduce the very inequalities NbS purports to resolve.

The sections that follow examine in detail how tenure insecurity, weak governance, and exclusionary practices undermine NbS outcomes, drawing on global evidence and regional case studies. Later sections explore the concept of "Good Enough Land Tenure Security" (GELTS) and propose practical pathways for integrating tenure reform into climate mitigation strategies.

Gender, Tenure Security, and Climate Resilience

Across the Global South, women are the backbone of rural economies. Securing women's land rights is not merely an issue of gender equity but an integral component of a legitimate, inclusive approach to climate mitigation that recognizes the unique role of IPLCs in managing land resources (Chesterman et al., 2022; Garnett et al., 2018; Gray et al., 2015; Oxfam International, 2021; Veit, 2019).

Women often make up nearly half or more of the agricultural workforce, yet they remain largely shut out of land governance and decision-making. The exclusion of women from land governance not only perpetuates gender inequality but also undermines the overall effectiveness of climate resilience strategies. Secure land rights for women are critical for ensuring that climate adaptation and mitigation efforts are both equitable and sustainable (Gray et al., 2015).

Discriminatory laws, social norms, and weak enforcement mean that women rarely have secure rights to the land they farm, even though their labor is central to food production and household resilience (FAO, 2021; Gumucio, 2024). This isn't just an issue of fairness; it's a major stumbling block for climate adaptation and mitigation.

When women do have secure land rights, the results are generally striking. Women with tenure rights are about 30% more likely to adopt practices like agroforestry, soil conservation, and drought-resistant crops. These are exactly the kinds of strategies that help communities adapt to climate change and store more carbon (Gumucio, 2024). In Senegal, for example, women's cooperatives that gained collective tenure rights ramped up carbon sequestration by 2.3 tCO_2/ha/year while also reducing desertification and improving food security (UN Women, 2022). Similar patterns are emerging in South Asia, where women's access to land has been linked to enhanced climate resilience in countries like Nepal and India, where women-led agricultural innovations have demonstrated significant climate adaptation benefits.

The economic and social ripple effects are just as important. In Kenya, Maasai women who secured land titles were able to access microloans, set up communal tree nurseries, and raise household incomes by nearly 25% (Chesterman, 2022). Women's involvement in land governance has also been shown to make communities less vulnerable to displacement during climate shocks, reducing those risks by 40% (Rights and Resources Initiative, 2021).

Still, most NbS projects are not set up to support women's land rights. In East Africa, only 15% of NbS initiatives include gender-specific tenure safeguards, and in Uganda, less than 10% of climate adaptation training reaches female smallholders (Gumucio, 2024; UN Women, 2022). These gaps keep women sitting on the sidelines, despite the evidence that their leadership and secure tenure have been shown to drive successful outcomes.

In Senegal's Thiès region, the Fédération des Femmes de la Diobass secured collective tenure and transformed degraded land through acacia-based agroforestry. Their work led to carbon sequestration rates 37% higher than nearby male-led projects, a 62% drop in soil erosion, and a boost in millet yields by over a ton per hectare, confirming assumptions about the wisdom of investing in women-led associations in natural resource management programming. The women also started a gum arabic cooperative that now reportedly generates $220,000 a year, proving that when women have secure rights, entire communities benefit (Gumucio, 2024).

As part of a paradigm shift, legal reforms must be encouraged to recognize women's rights to inherit and manage land. Nepal's community forestry programs have shown (Rights and Resources Initiative, 2024) the logic for doing so. Participatory mapping, as the work of Kenya's Samburu Women's Trust has shown, is a tool that can help women document and defend their claims to land (Claridge, 2018). So too, in ongoing project settings, climate finance clearly needs to be more gender-responsive. Costa Rica's FONAFIFO model shows what is possible when a significant share of funds goes directly to women-led initiatives (IUCN, 2020; Orozco Rubio and Sánchez, 2021).

In short, closing the gender gap in land tenure is not simply about justice. It's about supporting an evidence-based pathway to achieving more effective, resilient, and equitable climate solutions. In addition to securing land tenure, climate finance must be gender-responsive, ensuring that funds flow directly to women-led initiatives and gender-inclusive governance structures. This shift in financial models will enable women to lead the way in climate adaptation and mitigation efforts (Orozco Rubio & Sánchez, 2021).

Political and Economic Barriers to Land Tenure Reform

Addressing land tenure insecurity requires confronting the political and economic systems that enable dispossession, fail to recognize customary land rights, and weaken governance (USAID, 2013). Many governments still fail to recognize IPLC tenure systems, leaving communities vulnerable to elite capture and displacement (IWGIA, 2024).

For example, land grabbing facilitated by weak governance structures has led to forced displacement and conflicts over land in regions like the Amazon and Southeast Asia. These issues are compounded by systemic corruption that undermines land titling processes and delays justice for IPLCs (IWGIA, 2024). Corruption pervades the land sector and is closely linked to insecurity of land tenure. Corruption does not just delay legal processes – it actively reinforces power imbalances that prevent meaningful tenure reform, keeping land rights in the hands of a few elites while sidelining IPLCs and their rights.

Economic barriers also pose significant challenges. IPLCs often lack access to financial resources needed for legal processes such as land surveys or negotiations with external actors. In some cases, communities have resorted to piecemeal reclamation of territories due to prohibitive costs associated with securing formal titles (FAO, 2023).

Multilateral agencies and bilateral donors have long acknowledged land tenure insecurity as a critical barrier to climate and development goals. Yet, despite decades of rhetoric, donor strategies remain fragmented, technocratic, and divorced from the political realities of tenure reform. The unproductive reliance by the World Bank, in particular, on market-based land reform was noted over 20 years ago in an evidence-based indictment on the imposition of private entitlements and market-assisted land reforms (Tait, 2003).

Additionally, while the World Bank's 2003 Land Policy Report emphasized tenure security as a "foundation for development," 20 years later only 18% of World Bank land projects explicitly budget for legal recognition of customary rights (UNCCD, 2024). Similarly, the Global Donor Platform for Rural Development's Land Governance Working Group – comprising 28 members – has prioritized tenure since 2013. Yet evaluations reveal that 70% of donor-funded land initiatives still focus on technical fixes like GPS mapping or individual titling, rather than addressing corruption or power imbalances (Global Donor Platform, 2024). To be fair, the World Bank's efforts in Ethiopia discussed later in this chapter suggest that formalization across the board has not been a total failure.

The root issue lies in donors' preference for "manageable" technical interventions over contested structural reforms. To overcome these challenges, donors must pivot from market-driven to community-led, negotiated reforms where integrated land-use systems are prioritized. This requires tailored approaches that recognize local realities and promote legal pluralism and negotiated outcomes, as developed in more detail in Chapters 6 and 7.

A 2025 review of 120 climate projects found that 89% avoided funding legal challenges to corrupt land allocations or supporting IPLC-led litigation, fearing political backlash (JPIA, 2023). As one USAID land advisor conceded, "We're stuck in a loop – training communities to map lands they could lose tomorrow to a mining concession we helped approve" (Landesa, 2024).

Until donors confront their role in perpetuating elite capture and market-driven models, where compensation is often inadequate (Ghatak et al., 2013) tenure security will remain a checkbox, not a cornerstone, of development and sustainable climate action. To overcome the inefficiencies of donor fragmentation, it's essential to align donor strategies around community-led governance and legal pluralism, recognizing that land reforms must be tailored to local realities, not imposed from the outside.

Global Drivers of Tenure Insecurity: Where Does ESG Stand?

Environmental, Social, and Governance (ESG) frameworks have become a major force in global finance and corporate strategy, but their role remains hotly debated – especially in the United States, where ESG has become a political battleground. In recent years, Republican-led states have introduced and passed over 200 anti-ESG bills, framing ESG as a "woke" agenda or a threat to shareholder value (Forthomme, 2024; Hilson, 2024). This polarization stands in sharp contrast to Europe and much of the Global South, where ESG is widely viewed as synonymous with sustainability and a necessary response to climate and social risks (Forthomme, 2024). Despite the backlash, many corporations continue to integrate ESG into their business models, motivated by risk management, regulatory compliance, investor expectations, and the recognition that strong ESG performance can enhance reputation, lower capital costs, and drive long-term value (RPS Group, 2023).

ESG frameworks are now widely used to guide investment and corporate decision-making on climate, biodiversity, and the SDGs. IOs such as the United Nations, the Convention on Biological Diversity, and the Intergovernmental Science-Policy Platform on Biodiversity and Ecosystem Services highlight their significance (EcoActive ESG, 2024; Vu Phong Energy Group, 2024). ESG standards help companies and investors measure and report on ESG impacts and are increasingly linked to the SDGs as practical frameworks for sustainable development (CSR Training Institute, 2025; IRIS Business, 2024). Box 4.1 illustrates one case where ESG claims are recognized while local, Indigenous Peoples in Brazil nonetheless contest multi-billion dollar activities. While ESG standards appear to support sustainability, in practice, as the contested case illustrated in Box 4.1 suggests, ESG-purported projects that are state-licensed may fail to

Box 4.1 The ESG Paradox: Claimed Compliance, Yet Contested by Indigenous Communities

Despite the increasing prominence of ESG frameworks, many projects that claim to meet ESG standards remain highly contested, particularly by IPLCs. For example, the Autazes Potash Project in Amazonas State, Brazil, is claimed by Brazil Potash Corp. (2024) to operate under ESG principles. The company highlights its commitment to sustainable development, including reducing greenhouse gas emissions and supporting local employment. However, the project is highly contested by the Mura Indigenous communities, who allege that their land rights were not fully respected during consultations, and they fear environmental degradation such as water contamination and deforestation.

This controversy illustrates a significant paradox: The project has been licensed by the Amazonas Environmental Protection Institute (IPAAM) despite concerns raised by federal prosecutors and the local community (Associated Press, 2025). The company's commitment to ESG goals appears in tension with the failure to uphold Indigenous rights and secure land tenure, raising broader questions about the real-world impact of ESG compliance.

protect Indigenous land rights or secure effective, inclusive consultation that achieves genuine FPIC, leading to environmental harm and, in the worst cases, land dispossession.

Yet, the effectiveness of ESG frameworks in delivering real-world sustainability outcomes hinges on their ability to address systemic governance failures and land tenure insecurity – issues that remain poorly integrated into mainstream ESG strategies (Deloitte, 2024; Kopnina et al., 2024; Sustainalytics, 2025). Despite the proliferation of ESG reporting, biodiversity and land tenure remain underrepresented in ESG metrics, limiting their effectiveness for protecting ecosystems and supporting community rights (ScienceDirect, 2023; Sustainalytics, 2025). Recent regulatory trends, such as the Taskforce on Nature-related Financial Disclosures, underscore the urgent need for ESG frameworks to integrate biodiversity and land tenure risks to avoid greenwashing and ensure real progress toward SDG and climate targets (Deloitte, 2024; IRIS Business, 2024).

Even with their prominence until recent years, ESG frameworks have often overlooked land tenure security and Indigenous rights, enabling greenwashing rather than driving substantive change in land governance. For instance, current ESG frameworks fail to assess risks related to violations of Indigenous rights, which can lead to adverse consequences for IPLCs and undermine sustainable practices (Birss & Gualinga, 2022).

In summary, while ESG frameworks are increasingly central to global sustainability efforts, their transformative potential will remain unrealized unless they directly address the foundational issues of land tenure security and governance integrity (Blackman et al., 2017).

ESG Investment, Land, and Carbon in the Global South

Despite political polarization, ESG-driven investment in land and carbon projects in the Global South is growing rapidly. The voluntary carbon market is projected to reach $50 billion by 2030, with the majority of new projects located in Africa, Latin America, and Southeast Asia, according to Trove Research & International Emissions Trading Association (2023). These projects – ranging from REDD+ forest conservation to renewable energy and agroforestry – are often promoted as win-wins for climate, biodiversity, and local livelihoods. For example, Uganda's Nyamwamba II hydropower project and its PES scheme demonstrate how ESG-aligned finance can support both sustainable land management and community resilience (Mlotshwa, 2023).

However, the promise of ESG in the Global South is often undermined by weak land tenure systems, fragmented legal frameworks, and corruption. As the UNCCD highlights, every $1 invested in land tenure security can generate $18 in social and economic returns by reducing disputes and enabling sustainable investment (UNCCD, 2024). Yet, fewer than 10% of ESG-aligned carbon projects in Africa and Southeast Asia explicitly budget for tenure regularization, leaving communities vulnerable to displacement and elite capture (Arko, 2024). The result is that, while ESG investment is helping to scale climate finance, it too often replicates colonial patterns of exclusion and land dispossession unless robust tenure safeguards are embedded from the outset (Anthesis Group, 2024; UNCCD, 2024).

The Governance Gap: When ESG Replicates Colonial Patterns

In Brazil's Cerrado, ESG-driven agribusiness expansion has contributed to illegal land grabs and deforestation, enabled by corruption and fraudulent land titles. Major corporations and investors – including Bunge Limited, Harvard University, and Teachers Insurance and Annuity Association of America (TIAA) – have acquired vast tracts of land, often at the expense of local communities and Indigenous Peoples. Despite Brazil's formal ESG commitments, weak enforcement of land laws has allowed these exploitative dynamics to persist, turning the Cerrado into a hotspot for land speculation and violence against environmental defenders (Pitta et al., 2023; Transparency International, 2024a).

In Vietnam, ESG-aligned investors in the timber sector have sourced an estimated 30% of their imports from illegal logging in Cambodia's protected areas, facilitated by bribery and corruption. This illicit trade not only accelerates deforestation but also displaces Indigenous communities, who lack legal recourse to defend their territories (Transparency International, 2024b). This is similar to displacement of the Wiwa in Colombia (The Guardian, 2024). Meanwhile, corruption remains pervasive in the DRC, ranking it among the world's most corrupt countries (Trading Economics, 2024), framing all official and private sector activities in DRC (Mudinga & Wakenge, 2021; Rights and Resources Initiative, 2020).

The DRC's Virunga-Kinshasa Green Corridor, discussed in the following section, meanwhile, illustrates the risks of ambitious conservation projects that proceed without first securing land tenure (Greenpeace Africa, 2025). Despite attracting hundreds of millions in climate finance and being premised on "partnerships with communities," the project has yet to provide a clear, feasible blueprint for resolving entrenched tenure insecurity, human rights abuses, and ongoing conflict – particularly in eastern DRC, where over 2.4 million people were displaced in the first half of 2024 (Internal Displacement Monitoring Centre, 2024; Mahonghol, 2024; Mudinga & Wakenge, 2021). As will be discussed, without legal recognition of customary tenure and meaningful consultation, such projects risk deepening tensions and replicating colonial patterns of exclusion.

The take-home message is clear: ESG frameworks that do not prioritize land tenure security are likely to perpetuate cycles of dispossession and environmental injustice, undermining both climate and development goals. Secure, transparent, and inclusive land governance must be treated as a non-negotiable foundation for all ESG-aligned investment and conservation efforts (Anthesis Group, 2024; UNCCD, 2024).

Weak Governance and Corruption in Land and Climate Systems

Corruption does not just delay legal processes – it actively reinforces the power imbalances that prevent meaningful tenure reform. By sidelining IPLCs, these dynamics perpetuate systemic dispossession and exacerbate environmental degradation.

Corruption and weak governance are among the most pervasive and poorly addressed barriers to effective land-based climate mitigation in the Global South. Transparency International's (TI's) Corruption Perceptions Index (CPI) consistently reveals that countries with high levels of public-sector corruption face disproportionate challenges in implementing climate policies, safeguarding ecosystems, and protecting Indigenous and community land rights (Transparency International, 2024a). Yet corruption in land and climate governance is rarely confronted directly in global climate discourse, leaving interventions vulnerable to elite capture, misallocation of funds, systemic inequities, and programming failures.

The most useful perspective on the global context of forest crime and governance comes from the United Nations Office on Drugs and Crime's 2025 *Global Analysis on Crimes that Affect the Environment – Part 2a: Forest Crimes: Illegal Deforestation and Logging*. It offers the first comprehensive global overview of forest crime, highlighting the convergence between illegal deforestation, organized crime, and other illicit activities such as drug trafficking, illegal mining, and human trafficking (UNODC, 2025). The report underscores that forest crime infiltrates legal supply chains through fraudulent permits, bribery, and regulatory loopholes, often with the complicity of corrupt officials and legitimate businesses. These crimes are particularly acute in tropical regions, which account for 91% of global deforestation since 1990, and are facilitated by insecure land tenure, weak law enforcement, and lack of harmonized legal frameworks.

The UNODC stresses that closing regulatory loopholes, strengthening law enforcement, and empowering local communities are essential to combating the interconnected threats posed by forest crime to biodiversity, local livelihoods, and global climate stability (UNODC, 2025), with "calls to community-based actions," "building the capacity of forest community members," and "alternative livelihoods" for those vulnerable to illegal deforestation and logging. Interestingly, while the report cites the role of insecure land tenure and weak governance as drivers of forest crime, it fails to explicitly connect the dots that enhancing tenure security must become a top priority for all stakeholders to CMLSGS.

Land Grabs and Deforestation

In Brazil's Cerrado savannah, corruption contributes to illegal land grabs for agribusiness, driving deforestation and carbon emissions. Judicial bribery and fraudulent land titles have allowed elites to claim over 366,000 hectares – an area larger than Luxembourg – while silencing dissent through violence against environmental defenders. Despite Brazil's TI CPI score of 38/100 (2024a), weak enforcement of land laws has turned the Cerrado into a hotspot for land speculation, with deforestation rates to support European clothing and fashion giants' cotton purchases increasing by 43% in 2023 alone (Transparency International, 2024b). These patterns are replicated in Bahia as well. One report refers to the phenomenon in the Cerrado as an "ecocide," enabled by major institutions including Bunge Limited, Harvard University, and the retirement fund manager TIAA (Pitta et al., 2023). In all instances, be it Amazonia, the Cerrado, or the Gran Chaco, corrupt officials collude with agribusiness operators to bypass environmental regulations, undermine climate and biodiversity goals, and undermine community livelihoods while trampling on human rights.

Illegal Logging and Smuggling

Vietnam, with a TI CPI of 40/100, has become a hub for laundering illegally harvested Cambodian timber, facilitated by a network of bribes to customs officials and senior politicians. An estimated 30% of Vietnam's timber imports are sourced from illegal logging in Cambodia's protected areas, where corruption undermines the possibility of sustainable forest conservation efforts and carbon sequestration (Transparency International, 2024b). This illicit trade not only fuels deforestation but also displaces Indigenous communities, who lack legal recourse to defend their territories.

Weak Governance and Corruption in Land and Climate Systems

Corruption and weak governance are among the most pervasive and poorly addressed barriers to effective land-based climate mitigation in the Global South. TI's

CPI consistently reveals that countries with high levels of public-sector corruption face disproportionate challenges in implementing climate policies, safeguarding ecosystems, and protecting Indigenous and community land rights (Transparency International, 2024a). Yet corruption in land and climate governance is rarely confronted directly in global climate discourse, leaving interventions vulnerable to elite capture, misallocation of funds, and systemic inequities.

A global perspective is provided by the United Nations Office on Drugs and Crime's 2025 *Global Analysis on Crimes that Affect the Environment – Part 2a: Forest Crimes: Illegal Deforestation and Logging*. This report highlights how illegal deforestation, organized crime, and other illicit activities, such as drug trafficking and illegal mining, converge in forest-rich regions, particularly in the tropics, which account for 91% of global deforestation since 1990. These crimes are facilitated by insecure land tenure, weak law enforcement, and regulatory loopholes, often with the complicity of corrupt officials and businesses. The UNODC stresses that closing legal loopholes, strengthening enforcement, and empowering local communities are essential to combating these interconnected threats to biodiversity, livelihoods, and global climate stability (UNODC, 2025). However, the report also notes that while insecure tenure and weak governance are acknowledged as drivers of forest crime, the global response has yet to prioritize land tenure security as a fundamental solution.

Major Examples: Brazil, Cambodia, and the DRC

Brazil: Land Corruption, Deforestation, and Climate Failure

Brazil, home to the Amazon and Cerrado biomes, illustrates the destructive synergy between weak land governance, corruption, and climate failure. Despite progress in reducing deforestation in the late 2000s, the last decade has seen a dramatic reversal: Deforestation in the Amazon rose by 22% from 2020 to 2021 alone, reaching the highest rate in 15 years (Climate Action Tracker, 2022). Corruption and fraud are the main enabling mechanisms for land abuses, guaranteeing impunity for land grabbers and complicit public officials (Transparency International Brazil, 2024a). In the Cerrado, judicial bribery and fraudulent land titles have allowed elites to claim over 366,000 hectares – an area larger than Luxembourg – while silencing dissent through violence against environmental defenders (Transparency International, 2023). Insecure land rights and overlapping claims distort land-use decisions, encourage premature or destructive investment, and drive deforestation and emissions (Climate Policy Initiative, 2016). Brazil's CPI score of 38/100 in 2024 reflects persistent challenges, with land corruption undermining both climate and social justice goals. In late 2025, evidence suggests the situation is worsening (EIA, 2025).

Cambodia: Illegal Logging, Land Concessions, and Corruption

Cambodia remains a global hotspot for illegal logging and timber trafficking, with high risks attached to timber products originating from the country (Forest Trends,

2025). Economic Land Concessions (ELCs), intended for infrastructure and agriculture, are frequently used as tools for pseudo-legal timber extraction and laundering illegally harvested wood from natural forests, including protected areas (Global Initiative Against Transnational Organized Crime., 2021). Corruption is endemic in the allocation and oversight of ELCs, with concessions often awarded through political patronage and bribes, and enforcement undermined by collusion between business elites and state officials (Amnesty International, 2022). Indigenous and local communities face coercion, displacement, and violence for resisting illegal logging, while rare species such as Siamese rosewood are targeted for illegal export. Despite official bans and international attention, weak enforcement and persistent corruption continue to drive deforestation and biodiversity loss. Cambodia faces the paradoxical situation in which 41% of its surface has nature reserve status, but the country boasts one of the highest deforestation rates in the world, leading to the conclusion that protected areas that are not community-led fail to benefit either people or nature.

Democratic Republic of the Congo: Corruption, Land Tenure Insecurity, and the Paradox of Ambitious Conservation

The Democratic Republic of the Congo (DRC) is home to around 60% of the Congo Basin rainforest – the world's second-largest tropical forest – which absorbs roughly 1.5 billion tons of CO_2 annually and stores an estimated 27–30 billion tons of carbon (Global Initiative Against Transnational Organized Crime, 2021; Global Witness, 2024). Despite ambitious commitments to sustainable forest governance, the DRC's progress is persistently undermined by entrenched corruption, impunity, and weak enforcement (Global Witness, 2022; Transparency International, 2024a). Over 12 million hectares have been illegally allocated for logging since a 2002 moratorium, and nearly 90% of logging is estimated to be illegal (Forest Trends, 2022; Hance, 2014). The country's persistent governance failures are reflected in its 2024 CPI score of 20/100, ranking 162nd out of 180 countries (Transparency International, 2024a). This is despite the country receiving $500 million in forest protection aid (Global Initiative Against Transnational Organized Crime, 2021).

The DRC continues to experience one of the world's highest deforestation rates, losing over a million hectares of tree cover in 2023 alone (Global Witness, 2024). A 2022 government review found that many logging concessions had been allocated in breach of national law, often through opaque practices and influence peddling. Attempts to suspend illegal concessions were quickly reversed by court order, returning two million hectares of rainforest to commercial exploitation within months (Global Witness, 2024).

Research consistently shows that insecure land tenure is a structural driver of both environmental degradation and violent conflict in the DRC, especially in the east. The overlap and contestation between customary, informal, and statutory land tenure systems have upended livelihoods and security, enabled elite capture, and fueled violent disputes, particularly in areas with weak state presence and ongoing armed conflict (International Alert, 2012; Mudinga & Wakenge, 2022; Rights and

Resources Initiative, 2020). The failure of land reform has been attributed to an "overly technical vision of responses on the ground," rather than the deep structural changes required to address the political economy of land (Mudinga & Wakenge, 2022).

This paradox is thrown into sharp relief by the DRC's audacious plan to establish the Virunga-Kinshasa Green Corridor, which aspires to be the world's largest forest reserve project, covering 540,000 km², an area the size of France (World Economic Forum, 2025). The project is premised on "partnerships with communities that integrate conservation and restoration with green economic development." However, the planning process has yet to provide a clear, feasible blueprint for addressing the deeply rooted issues of land tenure insecurity, human rights abuses, and ongoing conflict – particularly in eastern DRC, where displacement and violence are most acute (Internal Displacement Monitoring Centre, 2024). In the first half of 2024 alone, conflict in eastern DRC – including provinces traversed by the Green Corridor – caused more than 2.4 million internal displacements, with land and resource control at the center of these crises (Internal Displacement Monitoring Centre, 2024).

The Green Corridor's success hinges on negotiated land-use agreements with communities. Yet in eastern DRC, armed groups like M23 control vast territories, and 90% of the land is governed by customary systems lacking legal recognition, leaving communities vulnerable to dispossession by elites or armed actors (Mudinga & Wakenge, 2022). Despite assurances of community inclusion, Indigenous Pygmy leaders have reported no prior consultation, warning the project risks encroaching on sacred lands and deepening tensions. Greenpeace Africa has cautiously welcomed the ambition but cautions that obtaining FPIC will be a major challenge, as the initiative's scale clearly exceeds the capacity of DRC's under-resourced conservation agencies and fails to address the corruption enabling industrial logging in protected areas (Global Witness, 2024). If illegal logging occurs in the DRC's protected areas, how will it be prevented in forests under customary community tenure rights?

The DRC's experience demonstrates that without confronting entrenched corruption, formalizing land tenure, and ensuring meaningful community participation, even the most ambitious conservation projects risk exacerbating existing conflicts and undermining both environmental and social outcomes (Global Witness, 2024; Mudinga & Wakenge, 2022; Transparency Internationl, 2024a).

Additional Examples: Tenure Insecurity for Pastoral Peoples in East Africa

Kenya

Pastoralist communities in counties such as Narok, Kajiado, Lamu, and Tana River face mounting land tenure insecurity as land titling, privatization, and agribusiness expansion fragment traditional grazing lands. The lack of legal protection for customary rights exposes these communities to land grabs and forced evictions, undermining both livelihoods and climate adaptation strategies (Holden & Ghebru, 2016). Although Kenya's Community Land Act (Community Land Act, No. 27

of 2016, Revised Edition 2022, Laws of Kenya, Cap. 287) was intended to secure collective rights, implementation remains slow and uneven, leaving millions vulnerable as external investment and state-backed development projects intensify competition for land (Otieno et al., 2024). Meanwhile, Kenya has committed to restoring 5.1 million hectares of degraded landscapes by 2030 through the African Forest Landscape Restoration Initiative (AFR100) and up to 10.6 million hectares by 2032 nationally. As institutional readiness in arid and semi-arid lands (ASALs) where land degradation is most severe is lacking (Chomba et al., 2021), Kenya's pastoralist communities face heightened challenges.

Uganda

In Uganda's Karamoja region, over 95% of land is held under customary tenure, but weak legal frameworks and government-backed land acquisitions have led to widespread dispossession and displacement (International Land Coalition, 2020). Large-scale agricultural and mining projects, often promoted as development, have eroded communal land rights, disrupted access to water and grazing, and increased vulnerability to food insecurity (Karamoja Resilience Support Unit, 2021). The absence of formal land rights excludes communities from climate finance and deepens poverty, while land conflicts remain unresolved due to overlapping statutory and customary systems (Seddon et al., 2020).

Tanzania

Maasai pastoralists in northern Tanzania have faced repeated waves of forced eviction and land dispossession for conservation and commercial agriculture since the 1970s (Amnesty International, 2023; Parkipuny, 1975). Despite some legal victories, the lack of formal recognition for customary land rights leaves Maasai communities highly vulnerable. Displacement and loss of access to grazing and water have led to severe hardship and the erosion of traditional livelihoods (Association for Law and Advocacy for Pastoralists (ALAPA), 2014; Behnke et al., 1993; Brown & Hughes, 2017).

Positive Models: Rights-Based Tenure Reform

Despite the many challenges detailed in this chapter, there are promising examples of negotiated, rights-based tenure reform that offer lessons for the region and beyond. In Kenya, the 2016 Community Land Act marked a historic step by enabling Indigenous and pastoralist communities to formally register and collectively own their communal lands. This law has allowed dozens of communities to secure legal title to millions of hectares, strengthening local governance, reducing conflict, and providing a legal foundation for negotiating with government and private investors (Alden Wily, 2018; Namati, 2024; Tenure Facility, 2022–2027).

In Ethiopia, a massive rural land certification program has issued more than 20 million certificates to smallholder households, including joint titles for women, at a fraction of the cost of similar efforts elsewhere. Rigorous evaluations show that certification has increased tenure security, boosted investment in soil and water conservation, and improved livelihoods, all while being implemented through decentralized, community-based land administration committees (Deininger et al., 2011). These cases demonstrate that when tenure reform is grounded in local participation and legal pluralism, it can deliver real social and environmental benefits, providing a counterpoint to the failures of top-down, technocratic approaches.

Land Tenure Security, Climate Justice, and Tackling Paternalistic Environmentalism

If you look across the Global South, it's hard to escape the fact that land tenure insecurity is inarguably a fundamental barrier to effective climate mitigation in the land sector. The evidence is overwhelming: Wherever land rights are weak or contested – whether in the DRC's forests, Brazil's Cerrado, Cambodia's timberlands, or the rangelands of East Africa – carbon sinks are lost, communities are sidelined, and even the best-intentioned climate or conservation projects get bogged down in conflict, elite capture, or outright failure. That said, it's just as clear that tenure insecurity is only one piece of a much bigger puzzle.

Corruption, for example, is a constant threat running through these stories, undermining land reform, distorting project incentives, and making it nearly impossible for communities to defend their rights or benefit from climate finance. But it doesn't stop there. Many IPLCs still lack comparable analytical capacity and negotiating power to effectively negotiate with governments and private sector actors on their own terms. The negotiation "space" itself is often narrow or non-existent, with top-down planning and donor-driven timelines leaving little room for appropriate, locally led solutions. Even where legal reforms or technical fixes are attempted, they tend to founder on the rocks of political economy: overlapping land systems, administrative opacity, and the political and participatory marginalization of IPLCs.

The mismatch between customary tenure systems and state laws has created legal confusion and land tenure insecurity. As a result, IPLCs often find themselves in conflict with both state actors and corporate interests, unable to assert their rights or access climate finance due to inconsistent land governance frameworks.

Without a negotiated model for land-use planning – one that starts with GELTS and builds up from there – projects like the aforementioned DRC Green Corridor are likely to reproduce the same failures we've seen elsewhere. The same goes for the ESG and climate finance space. If the underlying systems that shape how land rights are recognized, contested, and defended aren't fixed, then all the carbon markets, REDD+ schemes, and NbS in the world won't deliver lasting results.

So yes, land tenure insecurity is a central challenge, and it needs to be addressed head-on. But it's not the only priority. This book argues that we need a broader paradigm shift – one that tackles the political economy of land governance, builds IPLC capacity, expands negotiation space, and roots out corruption as part of the climate agenda. Chapters 6 and 7 will dig deeper into what this shift looks like in practice and how the different pieces – tenure, finance, governance, and local empowerment – can finally be integrated into a model that works for people, climate, and biodiversity alike.

In the end, there's no shortcut. To effectively address climate mitigation in the land sector, especially in the Global South, it will require more than just technical fixes or market adjustments. It will demand real power-sharing, genuine recognition of land rights, and a commitment to confronting the complex realities of politics, capacity, and historical context on the ground.

Land tenure security must be recognized as a foundational element. Without overcoming the political, economic, and social barriers to secure land rights, even the most well-intentioned projects will fall short of their goals and fail to deliver the sustainable outcomes needed to achieve long-term success.

References

Agrawal, A., & Gibson, C. C. (1999). Enchantment and disenchantment: The role of community in natural resource conservation. *World Development*, 27(4), 629–649.

Alden Wily, L. (2018). Collective land ownership in the 21st century: Overview of global trends. *Land*, 7(2), 68.

Amnesty International. (2022). *Land and Economic, Social and Cultural Rights: Observations on the Draft General Comment by the UN Committee on Economic, Social and Cultural Rights*. London: Amnesty International Ltd. https://www.amnesty.org/ar/wp-content/uploads/2022/11/IOR4046122021ENGLISH.pdf

Amnesty International. (2023). *"We Have Lost Everything": Forced Evictions of the Maasai in Loliondo, Tanzania*. London: Amnesty International. AFR 56/6841/2023.

Anthesis Group. (2024, December 12). Biodiversity, social, and climate considerations for corporate sustainability. https://www.anthesisgroup.com/insights/biodiversity-social-and-climate-considerations-for-corporate-sustainability/

African Land Policy Centre (ALAPA). 2014. Land Grabbing: Investments & Indigenous Peoples' Rights to Land and Resources. *IWGIA*, October 2014.

Arko, T. (2024, October 30). Carbon markets and the new scramble for African land. *Review of African Political Economy. (ROAPE)*. https://roape.net/2024/10/30/carbon-markets-and-the-new-scramble-for-african-land/

Associated Press. (2024). Guatemala must halt nickel mine and recognize Indigenous land rights, regional court rules. *AP News*. https://apnews.com/article/346b0c882d4ee52954a7bd32c81202a5

Associated Press. (2025). Potassium mining project in Brazil's Amazon rainforest divides Indigenous tribe. https://apnews.com/article/2a4dee3774950fb1b11c4f4836c94519

Behnke, R. H., Scoones, I., & Kerven, C. (1993). Range ecology at disequilibrium: New models of natural variability and pastoral adaptation in African savannas. *ODI Natural Resource Perspectives, No. 3*. Overseas Development Institute.

Birss, M., & Gualinga, E. S. (2022). Without indigenous rights it's not sustainable: Why corporate respect for Indigenous rights is an essential component of sustainable finance. *ResearchGate*. https://www.researchgate.net/publication/365771112_Without_Indigenous_Rights_It's_not_Sustainable_Why_Corporate_Respect_for_Indigenous_Rights_is_an_Essential_Component_of_Sustainable_Finance

Blackman, A., Corral, L., Lima, E. S., & Asner, G. P. (2017). Titling indigenous communities protects forests in the Peruvian Amazon. *Proceedings of the National Academy of Sciences*, 114(16), 4123–4128.

Brazil Potash Corp. (2024). Autazes potash project – ESG commitment. https://ir.brazilpotash.com/news-events/press-releases/detail/5/brazil-potash-receives-mine-installation-license-to-start-project-construction-of-the-autazes-potash-project-in-brazil

Brown, M., & Hughes, A. (2017, March 20–24). Is land tenure "secure enough" in rural Rwanda? Paper presented at the 2017 World Bank Conference on Land and Poverty, Washington, DC.
Chesterman, S. (2022). *Securing Land Rights for Women: Progress and Challenges in Kenya*. International Institute for Environment and Development (IIED). https://pubs.iied.org/21266g
Chesterman, S., Bailey, A., & Paez Valencia, A.M. (2022). Securing land tenure for women and girls to promote climate change resilience. *CGIAR*. https://www.cgiar.org/research/publication/securing-land-tenure-women-girls-climate-change-resilience/
Chomba, S., Mwangi, A. G., & Karanja, F. K. (2021). Forest landscape restoration in Kenya: Progress, challenges, and policy implications. *Forest Policy and Economics*, 129, 102506.
Claridge, L. (2018). Litigation as a tool for community empowerment: The case of Kenya's Ogiek. *Erasmus Law Review*, 11(1), 57–66.
Climate Action Tracker. (2022). *State of Climate Action 2022*. Systems Change Lab, Bezos Earth Fund, Climate Analytics, ClimateWorks Foundation, NewClimate Institute, and World Resources Institute.
Climate Policy Initiative. (2016). *Unleashing the Potential of Renewables in Brazil*. Climate Policy Initiative & Pontifical Catholic University of Rio de Janeiro (PUC-Rio).
COICA. (2020). *Indigenous Peoples Statement about the Amazon in the Context of the Pandemic and Climate Crisis: Women, Territory and Climate Change*. December 23, 2020. COICA.
Community Land Act, No. 27 of 2016 (Revised Edition 2022), Laws of Kenya, Cap. 287. Revised by 24th Annual Supplement (Legal Notice 221 of 2023) as at 31 December 2022. Published in Kenya Gazette Vol. CXVIII—No. 107 on 9 September 2016. Assented to on 31 August 2016. Commenced on 21 September 2016.
CSR Training Institute. (2025, January 14). ESG & SDGs: A powerful connection for profitable & sustainable businesses. https://www.csrtraininginstitute.com/knowledge-centre/articles/esg-sdgs-a-powerful-connection-for-profitable-sustainable-businesses/
Deininger, K., Selod, H., & Burns, A. (2011). *The Land Governance Assessment Framework: Identifying and Monitoring Good Practice in the Land Sector*. Washington, DC: World Bank.
Deloitte. (2024). Nature-related financial disclosures: The new frontier for ESG risks. https://www2.deloitte.com/us/en/pages/audit/articles/sec-climate-disclosure-guidance.html
Dooley, K., Keith, H., Larson, A., Catacora-Vargas, G., Carton, W., Christiansen, K. L., Enokenwa Baa, O., Frechette, A., Hugh, S., Ivetic, N., Lim, L. C., Lund, J. F., Luqman, M., Mackey, B., Monterroso, I., Ojha, H., Perfecto, I., Riamit, K., Robiou du Pont, Y., & Young, V. (2022). *The Land Gap Report. Melbourne, Australia: Melbourne Climate Futures*. The University of Melbourne. Available at: www.landgap.org. or https://research-repository.griffith.edu.au/items/1caf586a-ef16-4ef2-8d05-069ce189b462
EcoActive ESG. (2024). Integrating SDGs into your ESG reporting: A pathway to global impact. https://ecoactivetech.com/integrating-sdgs-into-your-esg-reporting-a-pathway-to-global-impact/
Environmental Investigation Agency (EIA). (2025, April 3). Illegal Logging Destroying the Brazilian Amazon Feeds EU and U.S. Markets. Retrieved from https://eia.org/press-releases/tricks-traders-and-trees/
Food and Agriculture Organization of the United Nations (FAO). (2021). *Realizing Women's Rights to Land in the Law: A Guide for Reporting on SDG Indicator 5.a.2*. https://openknowledge.fao.org/server/api/core/bitstreams/1aa1ae53-517f-4161-94da-99ce623a175d/content
Food and Agriculture Organization of the United Nations (FAO). (2023). Indigenous Peoples and tenure. https://www.fao.org/tenure/resources/collections/indigenouspeoplesandtenure/en/

Forest Peoples Programme, International Indigenous Forum on Biodiversity, Indigenous Women's Biodiversity Network, Centres of Distinction on Indigenous and Local Knowledge, & Secretariat of the Convention on Biological Diversity. (2020). *Local Biodiversity Outlooks 2: The Contributions of Indigenous Peoples and Local Communities to the Implementation of the Strategic Plan for Biodiversity 2011–2020 and to Renewing Nature and Cultures*. Moreton-in-Marsh: Forest Peoples Programme. https://www.cbd.int/gbo/gbo5/publication/lbo-2-en.pdf

Forest Trends. (2022). Timber legality risk dashboard: Democratic republic of the Congo. https://www.forest-trends.org/wp-content/uploads/2022/01/Democratic-Republic-of-the-Congo-Timber-Legality-Risk-Dashboard-IDAT-Risk.pdf

Forest Trends. (2025). *Illegal Deforestation and Associated Trade (IDAT) Risk: Cambodia Country Profile and Global Dashboard Update*. Washington, DC: Forest Trends. https://www.forest-trends.org/idat/

Forthomme, C. (2024). Why ESG is a political battleground in America. *Impakter*. https://impakter.com/why-esg-is-a-political-battleground-in-the-us/

Garcia O'Donnell, A. (2024). Green sacrifice zones: The justice implications of the green transition. *TASC*. tasc.ie. https://www.tasc.ie/blog/2024/02/06/green-sacrifice-zones-the-justice-implications-of/

Garnett, S. T., Burgess, N. D., Fa, J. E., Fernández-Llamazares, Á., Molnár, Z., Robinson, C. J., Watson, J. E. M., Zander, K. K., Austin, B., Brondizio, E. S., Collier, N. F., Duncan, T., Ellis, E., Geyle, H., Jackson, M. V., Jonas, H., Malmer, P., McGowan, B., Sivongxay, A., & Leiper, I. (2018). A spatial overview of the global importance of Indigenous lands for conservation. *Nature Sustainability*, 1(6), 369–374. https://doi.org/10.1038/s41893-018-0100-6

Ghatak, M., Mitra, S., Mookherjee, D., & Nath, A. (2013). Land acquisition and compensation in Singur: What really happened? *Economic and Political Weekly*, 48(21), 32–44. https://www.jstor.org/stable/23527421

Global Donor Platform for Rural Development. (2024). *Annual Report 2024*. Bonn: Global Donor Platform for Rural Development. https://www.donorplatform.org/post/annual-report-2024/

Global Initiative Against Transnational Organized Crime. (2021, May). Deep-rooted interests: Licensing illicit logging in Guinea-Bissau. *Risk Bulletin*. https://globalinitiative.net/analysis/illicit-logging-guinea-bissau/

Global Witness. (2022). Crisis in the Congo: How corruption and poor forest governance undermine climate progress. *Global Witness*. https://globalwitness.org/en/campaigns/forests/crisis-in-the-congo/

Global Witness. (2024). Why is the Congo Basin – the world's largest forest carbon sink – at risk? https://globalwitness.org/en/campaigns/forests/why-is-the-congo-basin-the-worlds-largest-forest-carbon-sink-at-risk/

Gray, E., Veit, P. G., Altamirano, J. C., Ding, H., Rozwalka, P., Zuniga, I., Witkin, M., Borger, F. G., Pereda, P., Lucchesi, A., & Ussami, K. (2015). *The Economic Costs and Benefits of Securing Community Forest Tenure: Evidence from Brazil and Guatemala*. Washington, DC: World Resources Institute.

Greenpeace Africa. (2025). DRC green corridor project: Greenpeace Africa cautiously welcomes world's largest forest reserve, urges inclusive approach. https://www.greenpeace.org/africa/en/press/56965/drc-green-corridor-project-greenpeace-africa-cautiously-welcomes-worlds-largest-forest-reserve-urges-inclusive-approach/

The Guardian. (2024). "We saw our family members cut into pieces": How Colombia's Wiwa people have been forced from their mountain again. The Guardian. Published May 13, 2024.

Gumucio, T. (2024). *Women's Land Tenure Security as a Pathway to Climate Change Mitigation and Adaptation: Evidence Scan*. Seattle, WA: Landesa. https://www.landesa.org/wp-content/uploads/WomensTenure_ClimateChange_EvidenceScan_2024.pdf

Hance, J. (2014, April 8). Nearly 90 percent of logging in the DRC is illegal. *Mongabay*. https://news.mongabay.com/2014/04/nearly-90-percent-of-logging-in-the-drc-is-illegal/

Hilson, C. (2024). Climate change and the politicization of ESG in the US. *Frontiers in Political Science*, 6, 1332399. https://www.frontiersin.org/articles/10.3389/fpos.2024.1332399/full.
Holden, S., & Ghebru, H. (2016). The implications of "secure enough" tenure in Sub-Saharan Africa. *Agricultural Economics*, 47(6), 695–708. https://doi.org/10.1111/agec.12272.
Internal Displacement Monitoring Centre. (2024). Democratic republic of the Congo: Internal displacement report (June-August 2024). https://reliefweb.int/report/democratic-republic-congo/democratic-republic-congo-internal-displacement-report-26-provinces-june-august-2024
International Alert. (2012). *Land, Power and Identity: Roots of Violent Conflict in Eastern DRC*. London: International Alert.
International Panel of Experts on Sustainable Food Systems (IPES-Food). (2024). Land Squeeze: What is driving unprecedented pressures on farmland and what can be done to achieve equitable access to land? IPES-Food. Available at: https://ipes-food.org/report/land-squeeze/
International Land Coalition (ILC). (2020). Land Grabbing & Conflict. *International Land Coalition*. Retrieved from https://www.landcoalition.org/en/whats-at-stake/issues/landgrabbing/
International Work Group for Indigenous Affairs (IWGIA). (2024). *The Indigenous World 2024* (D. Mamo, Ed.). Copenhagen: IWGIA.
IRIS Business. (2024, October 8). How to align ESG goals with UN sustainable development goals SDGs. https://www.irisbusiness.com/how-to-align-esg-goals-with-un-sustainable-development-goals-sdgs/
IUCN. (2020). *Costa Rica's Payment for Environmental Services (PES): Gender Responsive Land Tenure and Climate Finance*. IUCN Global Gender Office. https://genderandenvironment.org/resource/c
Journal of Public International Affairs (JPIA). (2023). Climate finance and legal risk: How funders and implementers navigate litigation and political backlash. *Princeton University School of Public and International Affairs, Journal of Public International Affairs*, 4(2), 49–71.
Karamoja Resilience Support Unit. (2021). *Land and Conflict in Karamoja: Assessment of Land Use and Control in Karamoja Region*. Kampala: Karamoja Resilience Support Unit (KRSU)/USAID/Uganda, UK aid, Irish Aid. https://karamojaresilience.org/
Kopnina, H., Zhang, S. R., Anthony, S., Hassan, A., & Maroun, W. (2024). The inclusion of biodiversity into Environmental, Social, and Governance (ESG) framework: A strategic integration of ecocentric extinction accounting. *Journal of Environmental Management*, 351, 119808. https://doi.org/10.1016/j.jenvman.2023.119808
Landesa. (2024). *Land Rights in the Age of Green Investments: Lessons from Field Implementation*. Seattle: Landesa Rural Development Institute.
Larson, A. M., Frechette, A., Ojha, H., Lund, J. F., Monterroso, I., Riamit, K., & Baa, O. (2022). Chapter 4: Land rights of Indigenous Peoples and local communities. *The Land Gap Report*. https://landgap.org/downloads/2022/Land-Gap-Report_Chapter-4.pdf
Larson, A. M., & Soto, F. (2008). Decentralization of natural resource governance regimes. *Annual Review of Environment and Resources*, 33, 213–239.
Lind, J., Okenwa, D., & Scoones, I. (Eds.). (2020). *Land, Investment & Politics: Reconfiguring Eastern Africa's Pastoral Drylands*. James Currey.
Mahonghol, D. (2024, November 4). The Regional Security Imperative to Protect the Congo Basin. Africa Center for Strategic Studies. https://africacenter.org/publication/asb44en-regional-security-imperative-protect-congo-basin/
Massarella, K., Sallu, S. M., Ensor, J. E., & Marchant, R. (2018). REDD+, hype, hope and disappointment: The dynamics of expectations in conservation and development pilot projects. *World Development*, 109, 375–385. https://doi.org/10.1016/j.worlddev.2018.05.006
Masuda, Y. J., Kelly, A. C., Robinson, B. E., Holland, M. B., Bedford, C., Childress, M., Game, E. T., Ginsburg, C., Hilhorst, T., Lawry, S. W., Miteva, D. A., Musengezi, J., Naughton-Treves, L., Sunderlin, W. D., & Veit, P. (2020). How do practitioners characterize land tenure security? *Conservation Science and Practice*, 2(5), e186.

Mudinga, E. M., & Wakenge, C. I. (2021). *Land crisis and stakeholders' responses in the Democratic Republic of the Congo (Congo Research Briefs, Issue 9).* Conflict Research Group, Ghent University; Social Science Research Council; GiC Network. https://www.gicnetwork.be/land-crisis-and-stakeholders-responses-in-the-democratic-republic-of-the-congo-2/

Mlotshwa, W. (2023, June 16). Land through a climate lens. *ESG Investor.* https://esginvestor.net/author/wendy-mlotshwa/

Muroa, G. (2022). Land disputes and conflict resolution mechanisms in Papua New Guinea: A case study of the Enga Province. *Land Use Policy*, 112, 105861.

Namati. (2024, May 17). Namati Submission SD-Tool_template_ver06.0.docx. Submission to the UNFCCC Supervisory Body: Call for public input – Draft Article 6.4 sustainable development tool (ver. 06.0). Submitted by Rebecca Iwerks. Available at UNFCCC website.

Nepstad, D., Shimada, J., Arima, E., Bauch, S., & Carvalho, O. (2019). Scaling up jurisdictional sustainability in the Brazilian Amazon. *Philosophical Transactions of the Royal Society B: Biological Sciences*, 375(1794), 20190126.

Oakland Institute. (2016). *Miracle or Mirage? Manufacturing Hunger and Poverty in Ethiopia.*

Orozco Rubio, A. L., & Sánchez, R. (2021). Women and nature programme: Putting women at the heart of biodiversity finance in Costa Rica. *UNDP Biodiversity Finance Initiative (BIOFIN).* https://www.biofin.org/news-and-media/women-nature-programme-putting-women-heart-biodiversity-finance-costa-rica

Otieno, K., Lumumba, O., Odote, C., Akinyi, L., Wari, G., Ongesa, L. and Nassef, M. (2024). *Perceptions of land tenure security in pastoral areas in Marsabit, Kenya. London: Supporting Pastoralism and Agriculture in Recurrent and Protracted Crises (SPARC).* https://www.sparc-knowledge.org/publications-resources/perceptions-land-tenure-security-marsabit-kenya

Oxfam International. (2021). *Toward a More Equal Future: Tackling Inequality in Land Rights.* Oxford: Oxfam International. https://oxfamilibrary.openrepository.com/handle/10546/621164

Parkipuny, M.S. (1975), *Maasai Predicament Beyond Pastoralism,* Unpublished MA Thesis, University of Dar es Salaam.

Pitta, F., Paris, T.; Mendonça, M. L., & Madan, G. (2023). *Land Grabbing and Ecocide: How Bunge, TIAA, and Harvard Fuel the Destruction of the Brazilian Cerrado.* Friends of the Earth U.S., ActionAid USA, and the Brazilian Network for Social Justice and Human Rights. https://www.actionaidusa.org/publications/land-grabbing-and-ecocide-how-bunge-tiaa-and-harvard-fuel-the-destruction-of-the-destruction-of-the-brazilian-cerrado/

Project Drawdown. (2024). *Indigenous Peoples' Forest Tenure.* Project Drawdown. https://drawdown.org/solutions/indigenous-peoples-forest-tenure

Reyes-García, V., et al. (2022). Recognizing Indigenous Peoples' and local communities' rights and agency in the post-2020 biodiversity agenda. *Ambio*, 51, 84–92.

Rights and Resources Initiative. (2018). *A Global Baseline of Carbon Storage in Collective Lands: Indigenous and Local Community Contributions to Climate Change Mitigation.* Rights and Resources Initiative. https://rightsandresources.org/wp-content/uploads/2018/09/A-Global-Baseline_RRI_Sept-2018.pdf

Rights and Resources Initiative (RRI). (2020). Community land rights in the Democratic Republic of the Congo (DRC): A summary of 2020 analyses. https://rightsandresources.org/wp-content/uploads/Summary-of-2020-DRC-RRI-analyses_Final-EN-Formatted.pdf

Rights and Resources Initiative (RRI). (2021). *Closing the Gap: Implementing Indigenous and Community Land rights.*

Rights and Resources Initiative (RRI). (2022). *Who Owns the World's Land? A Global Baseline of Formally Recognized Indigenous and Community Land Rights*. Washington, DC: RRI.

Rights and Resources Initiative (RRI). (2024). *Who Owns the World's Land?* (2nd ed.). https://rightsandresources.org

Robinson, B. E., Holland, M. B., & Naughton-Treves, L. (2014). Does secure land tenure save forests? A meta-analysis of the relationship between land tenure and tropical deforestation. *Global Environmental Change*, 29, 281–293.1

Roe, S., Streck, C., Beach, R., Busch, J., Chapman, M., Engelmann, J., Franks, M., & Funk, J. (2021). Land-based measures to mitigate climate change: Potential and feasibility by country. *Global Change Biology*, 27(23), 6025–6058. https://doi.org/10.1111/gcb.15873

RPS Group. (2023). Building a business case for ESG: How to achieve stakeholder buy-in and demonstrate value. https://www.rpsgroup.com/insights/consulting-uki/building-a-business-case-for-esg-how-to-achieve-stakeholder-buy-in-and-demonstrate-value/

Scoones, I. (1995). Living with uncertainty: New directions in pastoral development in Africa. *Development and Change*, 26(1), 1–30. https://doi.org/10.1111/j.1467-7660.1995.tb00517

Seddon, N., Chausson, A., Berry, P., Girardin, C. A. J., Smith, A., & Turner, B. (2020). Understanding the value and limits of nature-based solutions to climate change and other global challenges. *Philosophical Transactions of the Royal Society B*, 375(1794), 20190120. https://doi.org/10.1098/rstb.2019.0120

Stevens, C., Winterbottom, R., Springer, J., & Reytar, K. (2017). *Securing Rights, Combating Climate Change: How Strengthening Community Forest Rights Mitigates Climate Change*. Washington, D.C.: World Resources Institute. Available at: https://rightsandresources.org/wp-content/uploads/Securing-Rights-Combating-Climate-Change.pdf

Sunderlin, W. D., de Sassi, C., Sills, E. O., Duchelle, A. E., Larson, A. M., Resosudarmo, I. A. P., & Huynh, T. B. (2018). Creating an appropriate tenure foundation for REDD+: The record to date and prospects for the future. *World Development*, 106, 376–392.

Sustainalytics. (2025). Addressing the drivers of biodiversity loss: Key solutions for companies and investors. https://www.sustainalytics.com/esg-research/resource/why-biodiversity-is-good-for-business

Tait, S. (2003). Property matters: Synergies and silences between land reform research & development policy. *Journal of Public and International Affairs*, 14, 215–239. Princeton University. https://jpia.princeton.edu/document/333

Tauli-Corpuz, V., Alcorn, J., & Molnar, A. (2020). *Cornered by Protected areas: Replacing "Fortress" Conservation with Rights-Based Approaches Helps Bring Justice for Indigenous Peoples and Local Communities, Reduces Conflict, and Enables Cost-Effective Conservation and Climate Action*. Washington, DC, USA: Rights and Resources Initiative.

Tenure Facility. (2022–2027). *Securing registered title to our lands*. https://thetenurefacility.org/projects/securing-registered-title-to-our-lands/

Trading Economics. (2024). Congo corruption index (transparency international). https://tradingeconomics.com/congo/corruption-index

Transparency International. (2023). *Tackling Land Corruption is Vital to Achieving Land Rights and the 2030 SDG Agenda: Policy Position on Land Corruption and the SDGs*.

Transparency International. (2024a). Brazil country page. https://www.transparency.org/en/countries/brazil

Transparency International. (2024b). Cambodia country page. https://www.transparency.org/en/countries/cambodia

Trove Research & International Emissions Trading Association (IETA). (2023). *Investment Trends and Outcomes in the Global Carbon Credit Market*. London: Trove Research.

Tseng, T. W. J., Robinson, B. E., Bellemare, M. F., BenYishay, A., Blackman, A., Boucher, T., Childress, M., Holland, M. B., Kroeger, T., Linkow, B., Diop, M., Naughton, L., Rudel, T., Sanjak, J., Shyamsundar, P., Veit, P., Sunderlin, W., Zhang, W., & Masuda, Y. J. (2021). Influence of land tenure interventions on human well-being and environmental outcomes. *Nature Sustainability*, 4, 242–251.

UNCCD. (2024). Why invest in land tenure security? https://www.unccd.int/sites/default/files/2024-12/DRAFT_LandTenureBusinessCase_FOR%20CONSULTATION%20COP16.pdf

UNODC. (2025, April). Forest crime converging with other types of organized crime, fueling harms to the environment, communities and economy. https://www.unodc.org/unodc/en/press/releases/2025/April/unodc_-new-study-shows-forest-crime-converging-with-other-types-of-organized-crime--fueling-harms-to-the-environment--communities--and-economy.html

UN Women. (2022). Empowering women through climate-resilient agriculture in West and Central Africa: Key results of UN Women's flagship programme. https://wrd.unwomen.org/sites/default/files/2022-04/EN%20Women%20in%20Climate%20Resilient%20Agriculture%20in%20WCA_0.pdf

USAID. (2013). Land tenure and property rights framework: USAID issue brief. *U.S. Agency for International Development.*

Veit, P. (2019). *Land Matters: How Securing Community Land Rights Can Slow Climate Change and Accelerate the Sustainable Development Goals*. World Resources Institute.

Vu Phong Energy Group. (2024). The connection between ESG and SDGS: Towards sustainable development. https://vuphong.com/the-connection-between-esg-and-sdgs/

Walker, S., Alix-Garcia, J., Bartlett, A., & Calder, A. (2025). Land tenure security and deforestation: Evidence from a framed field experiment in Uganda. *Journal of Environmental Economics and Management*, 131, Article 103137. https://doi.org/10.1016/j.jeem.2025.103137

World Economic Forum. (2025, January 22). The Democratic Republic of Congo to create the Earth's largest protected tropical forest reserve. https://www.weforum.org/stories/2025/01/congo-kivu-kinshasa-green-corridor/

World Resources Institute (WRI). (2017). *Securing Rights, Combating Climate Change: How Strengthening Community Forest Rights Mitigates Climate Change.*

5 Climate Mitigation Finance to the Global South

Pledges, Gaps, and the Credibility Crisis in Climate Finance

Any serious analysis of climate finance in the Global South must begin with the numbers: the pledges made, the funds actually disbursed, and the vast gaps between them. These figures are not just technical benchmarks; they reflect the credibility of climate commitments and the viability of mitigation strategies moving forward. Yet the credibility of mitigation strategies is deeply compromised by the widespread integrity problems in carbon offset markets, where credits often fail to represent genuine, equivalent climate benefits (Atal & Sylvan, 2025).

In 2022, developed countries finally met the long-standing $100 billion annual climate finance target set at COP15 in 2009, albeit two years late (OECD, 2024). Yet this milestone did little to restore confidence, given that the actual needs are many times greater. The New Collective Quantified Goal (NCQG) aims to mobilize $1.3 trillion annually by 2035 (UNFCCC, 2024), with the World Resources Institute (2025) estimating a need of $5 trillion annually by 2050 to keep the 1.5°C target within reach.

Despite growing ambition on paper, delivery remains weak. The World Bank (2024) reports that just $42.6 billion of a $300 billion target was disbursed in 2024, illustrating a persistent gap between rhetoric and reality. This failure is not merely fiscal – it signals structural dysfunction across the climate finance architecture.

The disparity in financial flows is compounded by inefficiencies in delivery. Funds are often funneled through multilateral institutions, national governments, and large international NGOs, many of which absorb significant portions of the financing in administrative overhead (Rights and Resources Initiative, 2022). This model sidelines the very communities most responsible for maintaining ecosystem integrity and most vulnerable to climate shocks: IPLCs.

Even if the NCQG target is met, it will not lead to meaningful climate outcomes unless funding is tied to STAM that are evidence-based, socially legitimate, and technically feasible. Simply scaling up flawed delivery models will reproduce existing inequalities and fail to deliver real mitigation. The IPCC (2019) warns that climate finance must embed local knowledge and governance systems to be effective. Without this, even large-scale financial mobilization risks becoming a well-funded failure.

Even *if* litigation were to generate billions of dollars in potential climate finance from major carbon-emitting companies based on recent scientific findings

DOI: 10.4324/9781003380320-6

of a "quantifiable link between fossil fuel companies and specific climate-related damages" (Callahan & Mankin, 2025), this potential funding source will generate limited benefits if it is simply rolled out to support the current dominant paradigm.

In short, the credibility crisis in climate finance is not about the amount of money alone, but about how it is allocated, who controls it, and whether it supports context-specific, sustainable outcomes. The problem is structural, not merely fiscal.

Beyond Intermediaries: Funding IPLC-Led Mitigation

A transformative shift in climate finance must prioritize bypassing intermediaries and channeling funds directly to IPLCs, who steward some of the most ecologically vital landscapes for climate mitigation (Garnett et al., 2018). This model minimizes overhead, enhances accountability, and enables performance-based funding tied to community-defined indicators such as biodiversity, forest health, and tenure security (RRI, 2020; WRI, 2025).

Clear principles for operationalizing this shift have been outlined by the Interlaken Group and RRI (2023), who call for rights-based community monitoring frameworks that strengthen human rights due diligence and accountability in land-based finance systems.

While some donors point to the $100 billion milestone met in 2022 (OECD, 2024) as progress, many Global South governments and civil society groups view it as confirmation of systemic inequity. The stark divide between actual disbursements and the $5 trillion annual need to meet the 1.5°C target (WRI, 2025) reflects fundamental dysfunctions in both delivery mechanisms and the prevailing paradigm. Even the ambitious $1.3 trillion NCQG goal (UNFCCC, 2024) risks failure if deployed through STAM strategies that fail the dual test of social legitimacy and technical feasibility.

As the IPCC (2019) emphasizes, climate finance must be embedded in local governance systems to be effective. Without a shift away from business-as-usual (BAU) models toward context-specific, rights-based approaches, climate finance will continue to underperform, reinforcing extractive dynamics rather than catalyzing transformation. Lack of agreement over the specifics of Article 6.4 implementation, due in large part to the strong critique by many IPLCs, national NGOs, and a number of international NGOs (see Lang, 2024), persists in the run up to COP30 in Brazil.

Climate Finance: Pledges, Disbursements, and Systemic Bottlenecks

The architecture of global climate finance continues to exhibit a stubborn gap between rhetorical ambition and practical delivery. While high-level pledges have grown more ambitious, actual disbursements remain sporadic, slow, and often misaligned with need.

The $100 billion annual climate finance target, first pledged in Copenhagen in 2009, was not met until 2022 – and even then, it fell short of restoring confidence in the broader system (OECD, 2024). In 2024, under the NCQG, donors pledged $300 billion, yet only $42.6 billion was disbursed – just 14% of the commitment (World Bank, 2024). Looking ahead, the NCQG sets a 2035 target of $1.3 trillion

Table 5.1 Climate Finance: Pledges versus Disbursements

Year	*Pledges (USD Billion)*	*Disbursements (USD Billion)*	*Notes*
2020	100	89.6	Initial $100B/year target was missed until 2022 (OECD, 2024).
2022	115.9	115.9	First-time pledges matched disbursements.
2024	300	42.6	COP29 target; delivery fell far short (World Bank, 2024).
2035 (Target)	1,300	N/A	The NCQG ambition – feasibility remains uncertain (UNFCCC, 2024).

Sources: OECD (2024); UNFCCC (2024); World Bank (2024).

per year (Alayza & Larsen, 2025; UNFCCC, 2024). But this ambition remains speculative unless delivery systems undergo significant reform.

The deeper issue lies in how climate finance is structured and deployed. Current funding flows overwhelmingly pass through national governments, development banks, and international NGOs, leaving IPLCs – the primary stewards of many ecologically critical landscapes – at the periphery. These intermediaries absorb substantial administrative costs and often lack accountability to the communities most affected by climate change (RRI, 2022).

The World Resources Institute (2025) estimates that meeting the 1.5°C climate goal will require $5 trillion annually by 2050, underscoring how existing targets – let alone disbursements – are vastly insufficient. But funding scale alone is not the issue. Without a shift away from BAU delivery models, even fully funded pledges will fall short. As the IPCC (2019) makes clear, finance must be embedded within governance structures that are participatory, context-specific, and rights-based.

Table 5.1 highlights the disjuncture between high-profile pledges and actual disbursements.

The Persistent Shortfall

While headline figures and high-level pledges dominate climate finance discourse, the persistent shortfall between commitments and actual delivery reflects deeper, structural problems that go far beyond simple fiscal constraints. Recent studies show that disbursements of climate-related official development assistance are consistently 40%–50% below the amounts committed, with data on actual flows often incomplete or impossible to trace to their intended recipients (SIPRI, 2025).

This chronic gap is not just a technical reporting issue; it is rooted in political, institutional, and structural barriers that prevent funds from reaching those most vulnerable to climate change. Key obstacles include risk-averse fund management, lack of transparency and accountability, complex and costly accreditation and application processes, and a tendency to favor large, multi-million-dollar programs over flexible, locally led initiatives (Brookings, 2024 ; Solidaridad Network, 2022).

As a result, much of the finance that is mobilized fails to reach frontline communities, with funds often absorbed by intermediaries or lost in bureaucratic cycles. Without addressing these underlying barriers, such as mismatched funding timelines, burdensome reporting, and the prioritization of donor-driven priorities over local realities, even the most ambitious climate finance targets will remain largely aspirational (SEI, 2020; SIPRTI, 2025; Solidaridad Network, 2022).

To move forward, the climate finance system must shed its dependency on centralized, technocratic gatekeeping. Delivering climate finance that meets appropriateness and feasibility criteria will require a fundamental rethinking of who receives funds, how success is measured, and whether current systems reflect the realities on the ground (Oliver Wyman, 2023).

In discussing climate finance, it's essential to recognize that the current system, with its heavy focus on carbon markets, has failed to account for the complexities of Integrated Natural Resource Management (INRM). The market-driven models of carbon offsets and voluntary carbon markets distort the logic of integrated socioecological systems, particularly in the Global South (Hamerkop, 2023; Hodel et al., 2024). These systems, which are rooted in the land-based practices of IPLCs, prioritize land tenure, biodiversity, and cultural rights, elements that are sidelined in the pursuit of carbon credits for marketing purposes.

Structural Drivers Through a Substantivist Economic Anthropology Lens

The dominance of market-centric logic in climate finance – predicated on rational actors pursuing profit maximization – clashes irreconcilably with the socio-cultural realities of land-use decision-making in the Global South. Choices about forests, fields, and grazing corridors are shaped by kinship obligations, risk mitigation, spiritual values, and historical trauma, factors largely invisible to neoclassical economic models (Hann, 2021; Nature Sustainability, 2024).

For example, in rural China, strong kinship networks have been shown to deter coercive land takings by prioritizing communal reciprocity over individual profit (Zhang & Zhao, 2014). Similarly, Indigenous Chong communities in Cambodia's Southern Cardamom region resisted carbon commodification projects that threatened ancestral resin-tapping traditions rooted in intergenerational knowledge sharing (Human Rights Watch, 2024). This mismatch illustrates Polanyi's warning: Treating land as a financial commodity inevitably destabilizes the social fabric that sustains it (Dale & Holmes, 2023; Nature Sustainability, 2024; Polanyi, 1944).

The heavy reliance on debt and equity (89% of climate finance) assumes governments trying to meet their NDCs, and local communities, will prioritize repayment over ecological stewardship (IIED, 2024; IPCC, 2022). However, this model clashes with substantivist realities where reciprocity-based resource sharing governs landscape management. For instance, in Zambia's Luangwa Valley, according to the project developer, carbon revenues are channeled into community-chosen projects rather than individual payments, reinforcing collective stewardship norms (BioCarbon Partners, 2025). Others claim this is an example of "carbon colonialism," as the community has turned its life upside down in exchange for receiving a share of the income generated by BCP from certificate trading, amounting to 150,000 kwacha per year, or approximately 5,100 euros, all of which must be spent on community projects." Whether this will prove to be sustainable or not, given that villagers may see the initiative as neither appropriate nor feasible to their circumstances, remains to be seen.

While governments often bear the debt burden, communities indirectly face austerity measures or redirected funds, exacerbating vulnerabilities (Buchner et al., 2023; IIED, 2024). This precludes achieving CMLSGS.

Sub-Saharan Africa's meager 1.3% share of global climate finance stems not from risk but from donors' inability to engage with informal land governance systems (Anozie, 2024; World Bank, 2024). In the Sahel, pastoralist corridors negotiated through customary law – not titling regimes – are mismatched with finance models requiring formal tenure as collateral (FAO & Global Land Alliance, 2025). This institutional dissonance perpetuates exclusion, as seen in Mauritania and Chad, where entrenched gender inequities intersect with climate vulnerabilities (LandLinks, 2024).

Toward a Re-Embedded Finance Paradigm

The solution lies not in tweaking market models but in re-embedding finance within the social relations that govern land use. This requires replacing carbon-ton targets with co-designed metrics reflecting local priorities, such as pollinator diversity or soil fertility cycles. Namibia's communal conservancies exemplify this approach, blending state and customary governance to ensure ecological and cultural integrity (WWF, 2023). By recognizing informal tenure systems, Mali's phased agroforestry grants – aligned with seasonal decision cycles – increased uptake by 40%, demonstrating the viability of institutional hybridity (IUCN, 2024).

Non-market incentives must align with existing reciprocity systems. The Global Environment Facility's Small Grants Program channels 92% of funds directly to communities, bypassing intermediaries and reinforcing self-determined stewardship. These models prioritize negotiated governance over standardized efficiency metrics, ensuring finance resonates with the substantive realities of land-use decision-making.

Reimagining Markets Through Governance and Rights

The erosion of corporate net-zero commitments by 2025, marked by delayed timelines and diluted ambitions, exposes a critical tension in climate mitigation strategies: Markets, while flawed, are not inherently incompatible with sustainability. The failure lies not in markets themselves but in their context-blind design, which prioritizes profit over governance and carbon metrics over community rights. Voluntary pledges and carbon commodification have faltered because they treat ecosystems as abstract financial assets, divorced from the socio-political realities of land stewardship in the Global South (Hanbury, 2024). Yet, this does not negate the potential for markets to contribute to sustainability restructured around *negotiated* governance, IPLC leadership, and context-specific feasibility.

The retreat of private sector net-zero rhetoric, exemplified by energy executives pushing timelines to 2070 (Bain & Company, 2025) and fossil fuel firms planning for peak oil demand in 2038 (JPT, 2023), reflects a systemic misalignment between market incentives and ecological imperatives. However, the solution is not to discard markets but to redefine their role.

Carbon markets could, in theory, mobilize finance for sustainable land use that minimizes deforestation, sequesters carbon, and conserves biodiversity only if reconfigured as tools to amplify IPLC-led stewardship rather than extractive intermediaries. For instance, the ENSURE framework's attempt to standardize NbS metrics (CEEW, 2025) struggles because it prioritizes investor returns over community-defined outcomes, illustrating the pitfalls of technocratic market design.

Achieving sustainable land use requires centering *appropriateness* and *feasibility* in policy frameworks. Appropriateness demands that interventions align with local ecological knowledge, tenure systems, and cultural values. Feasibility necessitates governance structures that secure land rights, ensure equitable benefit-sharing, and embed accountability (FAO, 2012).

The Kasigau Corridor REDD+ project's failures involving sexual harassment and disrupted pastoralist livelihoods (SOMO & KHRC, 2023) stem from imposing carbon logic on landscapes governed by kinship and seasonal grazing patterns. Conversely, Namibia's communal conservancies demonstrate that blending state and customary governance can enhance ecological and social outcomes when IPLCs control decision-making (World Wildlife Fund, n.d.).

The path forward lies in reimagining carbon markets as platforms for *negotiated carbon products* if they are to work for people and the planet. This means IPLCs will need to define acceptable terms, metrics, and benefit-sharing mechanisms. This requires:

1. **Tenure Security as a Prerequisite**: Markets cannot function equitably without legally recognized land rights. In the Sahel, pastoralist corridors negotiated through customary law clash with carbon projects requiring formal titles, perpetuating exclusion (FAO & Global Land Alliance, 2025). Securing collective tenure, as seen in Zambia's Luangwa Valley, enables communities to engage markets on their terms, channeling revenues into self-determined priorities like schools and clinics (BioCarbon Partners, 2025).
2. **Participatory Governance Structures**: Carbon transactions must be governed by tripartite platforms where IPLCs, states, and donors jointly set rules. For example, the proposed Tropical Forests Forever Facility (TFFF) discussed in detail in the next chapters risks elite capture without such structures, but could succeed with enforceable co-governance (Climate Change News, 2025). Similarly, the Architecture for REDD+ Transactions (ART) requires robust safeguards to prevent governments from sidelining IPLCs in jurisdictional programs (CEEW, 2025).
3. **Beyond Carbon-Centric Metrics**: Success must be measured by IPLC-defined indicators – biodiversity health, soil fertility, and intergenerational knowledge

transmission – not just carbon tons. Mali's phased agroforestry grants, aligned with seasonal decision cycles, increased uptake by 40% by prioritizing drought resilience over rigid sequestration targets (IUCN, 2024).

4. **Direct Finance and Risk Mitigation**: Bypassing intermediaries, as has been demonstrated by the Global Environment Facility's Small Grants Programme in limited cases, ensures 92% of funds reach communities (GEF SGP, 2024; World Bank, 2022). Meanwhile, blended finance models with advanced payments or insurance mechanisms, if systematically implemented, as proposed for emerging market and developing economies, can de-risk investments while protecting IPLCs from volatile carbon prices (Convergence & Climate Policy Initiative, 2024).

The private sector's wavering commitments underscore that markets alone cannot deliver sustainability. Yet, when anchored in rights and governance, they can complement public finance and community stewardship. For example, Climate Impact X auctions prioritize projects with tangible social co-benefits, leveraging buyer demand for legitimacy (Climate Impact X, 2022). Similarly, the Territorial Governance Facility partners with IPLCs in Amazonia and Mesoamerica to design carbon projects that align with territorial management plans (Ecosystem Marketplace, 2024).

In sum, the crisis of climate finance is not a failure of markets *per se* but of their disconnection from the lived realities of land use. By recentering markets on negotiated governance, tenure security, and IPLC agency, carbon finance can evolve from a tool of extraction to a catalyst for stewardship – provided it is subordinated to the broader goal of sustainable land systems. As the IPCC (2021) stresses, even $1.3 trillion annually will fail without governance systems that prioritize legitimacy over liquidity. The choice is not between markets or justice but between markets that replicate colonial dynamics and markets that serve as instruments of collective ecological renewal.

When Mitigation Steals From Adaptation: A Gendered Cost

The persistent imbalance between mitigation and adaptation finance is not just a technical or budgetary issue – it is fundamentally a matter of gender justice. Women in rural communities across the Global South shoulder the bulk of food production, managing between 60% and 80% of agricultural activities (FAO, 2023). Yet, when climate finance overwhelmingly favors mitigation projects, the urgent and tangible needs of these women often go unmet. For example, in Kenya's arid lands, women spend six to nine hours daily fetching water during droughts – a burden that could be dramatically reduced with targeted investments in rainwater harvesting and water infrastructure (Oliver Wyman, 2023). Despite this, carbon credit projects in the same region received more than twice the funding allocated to water access initiatives in 2023 (Carbon Brief, 2024; UN Women, 2024).

This disparity reflects a deeper problem: The metrics that drive funding decisions systematically favor mitigation projects because their outcomes, such as tons of carbon dioxide reduced, are easier to quantify. In contrast, the essential work women do in adaptation—saving seeds, regenerating soil, and responding to disasters—resists commodification and often goes invisible in climate finance frameworks and accounting (Eriksen et al., 2021). The consequences are profound. In Mozambique, for instance, women living near a large-scale carbon offset plantation reported spending four additional hours each day foraging for firewood, as mitigation projects displaced subsistence farming and reduced local resource access (ActionAid, 2023).

When adaptation is properly funded and designed with gender in mind, the results speak for themselves. Senegal's Devolved Climate Finance model, which requires that half of the village committee members be women, has achieved a remarkable 90% success rate in project implementation (Global Center on Adaptation, 2024). This success is no accident. It reflects how women's participation ensures that climate interventions address real, immediate needs and build lasting resilience, with Rwanda an example of *gender-transformative approaches that embed women's leadership and agency in climate decision-making* (Development Initiatives, 2024).

To move beyond this false choice between mitigation and adaptation, climate finance must be reoriented with gender-responsive priorities at its core. One promising approach is adopting a "Gender-Adjusted Cost of Delay" metric, which accounts for how climate inaction disproportionately increases women's unpaid care work (Grown et al., 2024).

For example, every million dollars not invested in flood-resistant infrastructure translates into tens of thousands of additional hours women spend fetching water or caring for affected family members in flood-prone areas (IPCC, 2022). Another innovative strategy is redirecting carbon tax revenues to support women-led adaptation cooperatives, as piloted successfully in Colombia's Amazonian departments (UNDP, 2024).

Finally, it is critical to decolonize how success is measured in climate finance. Rather than focusing narrowly on emissions reductions, indicators should include metrics such as the percentage of women with secure post-disaster livelihoods or reductions in girls' school dropout rates following adaptation investments (UNFCCC, 2023). This shift would recognize the full spectrum of climate resilience and justice, centering the lived realities of women who are often the first responders and long-term stewards of their communities' survival.

To state the obvious, gender-responsive finance must be mainstreamed across all mechanisms, for example, mandating women's participation in carbon market governance (UNDP, 2024), including their participation in negotiating land-use plans for all mitigation projects. Ideally, adaptation actions and metrics such as drought resilience (Global Center on Adaptation, 2024) will habitually become embedded in all negotiated mitigation projects as well.

Compliance Carbon Markets and Article 6: The Same Problem, Institutionalized

While voluntary markets have drawn the most scrutiny, the compliance mechanisms under Article 6 of the Paris Agreement are now formalizing many of the same dynamics. Article 6.2 allows countries to trade emissions reductions bilaterally, while Article 6.4 establishes a UN-administered global carbon market. The official narrative – reaffirmed at COP29 – presents these tools as breakthroughs in international cooperation and finance mobilization. But for many Global South governments and civil society groups, Article 6 risks enshrining "carbon grabbing" as global climate policy.

At COP29, rules were finalized without adequate protections for Indigenous and local communities, raising fears that emissions reductions will be transacted without tenure reform, FPIC, or benefit-sharing (Belfer Center, Harvard Kennedy School, 2024). Akinwumi Adesina of the African Development Bank publicly criticized the $10-per-hectare offers extended to African countries for carbon sequestration, calling them exploitative and emblematic of broader inequities (Adesina, 2025).

Despite calls from Nicholas Stern, Sukdhev, who led The Economics of Ecosystems and Biodiversity, and others for carbon prices in the $50–$100/ton range (Carbon Pricing Leadership Coalition, 2019; High-Level Commission on Carbon Prices, 2017), the prevailing market rate remains below $7. This is insufficient to fund the full transaction costs of socially legitimate mitigation (CarbonCredits.com, 2025; World Bank, 2024). In short, Article 6 may enable the Global North to meet mitigation targets on paper while exporting climate burdens to the Global South through financialized but fundamentally extractive mechanisms with dubious integrity (Atal and Sylvan, 2025; Boell Foundation, 2023; Bretton Woods Project, 2024).

Systemic Barriers and Real-World Consequences in Climate Finance

The structural flaws of the climate finance system are not just bureaucratic abstractions. They have immediate, often devastating impacts on the communities they claim to serve. IPLCs, who steward vast carbon-rich ecosystems, remain largely locked out of direct funding by design. Complex accreditation processes, burdensome reporting requirements, and grant sizes tailored for large institutions systematically exclude grassroots groups. A 2022 survey found that only six out of 75 IPLC organizations had annual budgets over $1 million, with many struggling to meet donor demands for compliance with opaque frameworks (Cannon, 2022a). This exclusion is not incidental; it reflects a system optimized for intermediaries, not justice. Less than 1% of international climate finance reaches IPLCs directly, with the vast majority "evaporating" through layers of multilateral banks, government agencies, and large NGOs before it ever touches the ground (Euronews Green, 2023; Rainforest Foundation UK, 2024).

Market failures further compound these inequities. Carbon prices in voluntary markets continue to languish around $5–7/ton – far below the $50–$100/ton widely cited as necessary to fund equitable stewardship and robust safeguards (Goulder & Hafstead, 2022; Sylvera, 2023). As a result, projects are incentivized to prioritize cheap, low-integrity solutions over community well-being. The Kasigau Corridor REDD+ project in Kenya, for instance, has been lauded for its environmental achievements (Wildlife Works, 2024), but a 2023 investigation documented that revenues were diverted to international brokers while local women faced decision-making (SOMO & Kenya Human Rights Commission, 2023). When carbon is treated as a commodity rather than a commons, human rights and local priorities become collateral damage.

The situation is further destabilized by the recent withdrawal of major US banks, including JPMorgan, Bank of America, Citibank, Wells Fargo, Goldman Sachs, and Morgan Stanley, from the Net Zero Banking Alliance, a coalition formed to align finance with Paris Agreement goals (Heeb, 2025). These institutions, managing over $14 trillion in assets, cited political and antitrust pressures under the incoming Trump administration, which has vowed to dismantle ESG initiatives and reprioritize fossil fuels. This retreat undermines the credibility of voluntary climate alliances and shrinks the pool of capital available for sustainable land management projects in the Global South.

For CMLSGS and IPLCs, the implications are dire. With US banks deprioritizing climate-aligned lending, the already limited access to green finance for IPLCs and community-based projects is likely to shrink further. This will reinforce the dominance of multilateral development banks and large NGOs, which often impose top-down agendas and allocate only a fraction of resources to IPLCs themselves (Climate Investment Funds, 2024; Rainforest Foundation UK, 2024). The risk is that extractive industries and speculative investors will fill the vacuum, accelerating deforestation, land grabs, and the marginalization of IPLCs – just as seen in the DRC's oil auctions in peatlands (Cannon, 2022).

Donor dependency amplifies these risks. The abrupt freeze of USAID funding in 2025 sent shockwaves through the development world, forcing the agency to cease all activities and issue stop-work orders to nearly all funding recipients (Bhuyan, 2025; Earth.Org, 2025). Zambia's Luangwa Valley communities, once celebrated for channeling carbon revenues into schools and clinics, lost 70% of their budget overnight. To survive, they were forced into exploitative private equity deals that prioritized investor returns over ecological or social outcomes (BioCarbon Partners, 2025). This volatility exposes the fragility of models that tether climate action to shifting donor priorities and geopolitical whims rather than community sovereignty.

These dynamics are further entrenched by a lack of transparency and accountability among the major multilateral development banks that dominate climate finance. Multilater Development Banks (MDBs) often fail to disclose what is actually counted as climate finance, with Oxfam unable to verify up to 40% of World Bank claims in a single year (Bretton Woods Project, 2024). MDBs' reliance on loan-based finance, rather than grants or direct transfers, deepens debt burdens in the Global South and sidelines the just transition demanded by workers and frontline communities.

Together, these realities reveal a system that perpetuates extraction under a green veneer. Until finance is decoupled from intermediaries, anchored in tenure security, and governed by those living with the consequences, it will remain a tool of exclusion, not renewal. As Indigenous leaders have repeatedly pointed out, the barriers are not about IPLCs' capacity or willingness. They are about a donor-driven system that refuses to relinquish control (Euronews Green, 2023; Rainforest Foundation UK, 2024). Only a paradigm shift toward finance that is based on negotiated outcomes between financiers, governments, *and* IPLCs across the Global South will lead to sustainable outcomes.

Conservationists like Bhuyan are right to consider alternative approaches to financing, given the growing spectrum of urgent conservation priorities being faced, and that donor funding is no longer reliable. With USAID's folding under the Trump administration and Elon Musk's DOGE, along with cutbacks in the UK's FCDO budget and other EU countries such as Germany, Sweden, France, Belgium, and the Netherlands following suit, new funding sources for conservation and development are needed.

Hence, the turn to philanthropic or innovative financing approaches to pick up the slack. These approaches include flexible philanthropy, exemplified by Mackenzie Scott's approach to philanthropy that offers "gifts that are many times the norm in terms of size, totally unrestricted with respect to use and time constraint for expenditure, and virtually free of any reporting requirements"; participatory grantmaking; and blended finance models like Costa Rica's PES program that offer a "mix of rules and regulations (for example, it is forbidden to cut primary forest) and positive rewards that invite stakeholders to respond to incentives and disincentives" (Porras & Chacón-Cascante, 2018) or REDD+ programs such as the Verra-certified Chyulu Green Hills project in Kenya, where Abatable has reported that 90% of revenues from REDD+ carbon sales go to communities, which, through the Chyulu Hills Conservation Trust, decide how revenues will be "directly reinvested in field activities, including forest protection, biodiversity preservation and ecological restoration," and on "supporting local communities through school bursaries and healthcare and reducing the environmental impact of livestock farming."

While it is conventional wisdom that no one-size-fits-all approach to climate finance should be assumed, one glaring omission that this book argues for is acceptance of a universal principle – negotiated land-use agreements, where IPLC roles are the outcomes of informed decisions and negotiated terms. This is distinguished from reliance on external actors to define terms, which more often than not, IPLCs in the Global South will accept for lack of better options. This, however, will *not* mean they are either appropriate or feasible to induce sustainable outcomes. Only negotiation will raise the probability of designing interventions that are both appropriate and feasible, which in turn offer prospects for sustainability.

Namibia's Communal Conservancies: Lessons for Rights-Based Climate Finance

Namibia's communal conservancies, established over 35 years ago, demonstrate how climate finance can succeed when it prioritizes tenure security, direct community control, and ecological stewardship over extractive market logic. Recognized under the 1996 Nature Conservation Amendment Act, these conservancies blend statutory rights with customary governance, enabling communities to manage 20% of Namibia's land while retaining 90% of revenues from tourism and wildlife credits (Boudreaux, 2010; NACSO, 2024). This hybrid model has tripled elephant populations, revived desert lions, and generated $10 million annually for 227,802 conservancy members, proof that finance grounded in governance and rights outperforms carbon-centric approaches (Riehl et al., 2015). More importantly, Namibia's conservancy model shows that when conservation finance in the Global South focuses on tenure security, customary governance, and direct benefit-sharing, community-led conservation and climate benefits can follow.

While supportive of the role of tenure security and customary governance linked to finance, scaling the model faces limits: Few countries have Namibia's legal frameworks or tourism potential, and few donors support decades-long institution-building. Until climate finance prioritizes tenure reform and IPLC sovereignty over short-term metrics, Namibia's success will remain an exception, not a norm.

The Climate Finance Bottleneck: A Misguided Focus

The prevailing belief that a lack of climate finance is the primary bottleneck for climate mitigation in the Global South has led to misguided policies that focus disproportionately on mobilizing funding without addressing fundamental governance failures. While financing is undoubtedly crucial, increasing financial flows alone will not result in sustainable climate outcomes. Many conservation and climate investment projects have failed to deliver intended results, instead serving as vehicles for large organizations to secure continued funding with little accountability (Standing, 2024). Throwing money at climate mitigation challenges will do little if the STAM deployed are infeasible and unsustainable, as many failed efforts over the past 20 years of both climate and biodiversity conservation funding suggest.

Recent estimates from the World Economic Forum suggest that the global climate finance gap will reach $9 trillion by 2030, requiring an annual investment of $5 trillion (World Economic Forum, 2025). However, private sector financing alone cannot replace strong governance structures, transparent enforcement mechanisms, and effective land-use policies. Emerging economies such as Kenya, Indonesia, and Brazil face significant obstacles in securing meaningful climate investments due to a combination of weak institutional capacity, entrenched corruption, and land tenure disputes (Bhattacharya et al., 2020).

Greenwashing, Greenhushing, and Market Failures

Greenwashing – where companies exaggerate their sustainability commitments – and greenhushing – where they intentionally obscure their environmental efforts – are pervasive problems. These deceptive practices allow corporations to present themselves as environmentally responsible while continuing unsustainable activities. Current market-based mechanisms, including ESG frameworks and voluntary carbon markets, have not consistently produced meaningful emission reductions. Instead, they often serve as distractions from deeper systemic issues. For instance, studies have shown that high ESG-rated firms do not necessarily have lower carbon emissions, suggesting that ESG ratings may not effectively incentivize genuine environmental performance (Amel-Zadeh & Serafeim, 2018).

A notable example is the Brazilian Amazon, where multinational corporations continue to participate in deforestation-linked supply chains despite publicly pledging zero-deforestation policies. Companies like Cargill have been implicated in the destruction of ecosystems such as the Amazon, Grand Chaco, and Cerrado for the production of soy and beef, highlighting a disconnect between corporate commitments and actual practices (Human Rights Watch, 2021).

Similarly, carbon offset programs have faced criticism for their lack of integrity. Concerns have been raised that some methodologies approved by carbon standard setters are insufficiently rigorous, potentially leading to companies claiming emission reductions without actual environmental benefits (Schneider et al., 2024). Additionally, the improper use of voluntary carbon credits by companies for offsetting purposes has led to greenwashing allegations, undermining market demand confidence (ISDA, 2023).

Skepticism regarding the seriousness and integrity of corporate commitments to reduce GHG emissions, particularly Scope 3 emissions along their supply chains, has been fueled by studies indicating that a significant proportion of ESG funds may involve greenwashing (Timperly, 2021). For example, a report found that over 55% of ESG funds committed greenwashing through exaggerated claims, and 70% failed to meet their ESG targets (Boffo & Patalano, 2020).

The polarized views on corporations' actual intentions to genuinely mitigate Scope 3 emissions are evident among both strong proponents of robust follow-through with net-zero commitments and critics who argue that corporations have been gaming the system, deploying greenwashing or greenhushing to create the appearance of action when there is little substantiating evidence. This issue is particularly relevant to NbS and NCS as part of corporate carbon offsetting strategies, and it underscores the role that large international non-governmental organizations (BINGOs) and consulting firm partners play in supporting measurably high-integrity projects on the ground.

While not drawn from peer-reviewed research, Figure 5.1 circulated widely on LinkedIn feeds in early 2025 – produced by Data Driven Labs at the University of North Carolina – captures the growing skepticism toward corporate climate pledges, particularly regarding Scope 3 emissions. As gray literature, it should be interpreted as indicative rather than definitive. Yet the reactions it provoked – ranging from strong support to detailed rebuttals – underscore a broader legitimacy crisis in ESG reporting and the widespread perception that many net-zero commitments may be little more than greenwashing. It illustrates how public discourse is evolving in parallel with academic and policy critiques of voluntary carbon markets and corporate accountability frameworks. While ESG checks all the right boxes on paper, many doubt it delivers in practice (Perkins, 2024).

Potential Greenwashing Indicator	Among companies without a pledge or target (N = 645)	Among companies with target or pledge (N = 3997)
Company has no interim target	78%	29%
Company provides no scope 3 coverage	98%	69%
Company has no plan to achieve their emissions reduction target	27%	17%
Company plans for questionable use of carbon credits and offsets	26%	38%
Inventory doesn't cover appropriate gases for target	NA	11%
Company lobbies against climate policies	15%	9%
Company is not on track to achieve their target	NA	5.5%
Any potential greenwashing	NA	95%

Figure 5.1 Potential Corporate Greenwashing

The attribution of 95% of companies with potential for greenwashing was applauded by most participants in the thread and received pushback from others, with one commenter saying:

> The suggestion that there is "potential greenwashing" in 95% of cases with a target or pledge is apparently based on applying a logical "or" across all the other columns shown. But the two biggest contributors to that ("Company provides no scope 3 coverage" and "Company plans for questionable use of carbon credits and offsets") are both problematic in that regard. Scope 3 is hard, and many companies are working diligently on that without yet a deliverable available. Is it right to accuse those companies of greenwashing, "potential" or otherwise? Can we reasonably conclude today that the only reason a company wouldn't yet have scope 3 available is an intent to greenwash?[1]

The polarized views of corporations actual intentions to genuinely mitigate Scope 3 emissions based on evidence continue to be clearly on display among the strongest proponents of strong follow-through with Net Zero commitments on the one hand, and those arguing that corporations have been gaming the system deploying greenwashing to make the appearance of action when in fact there is little substantiating evidence (Atal & Sylvan, 2025). The issue is of relevance to NbS and NCS as part of corporations' carbon offsetting and the role that BINGO and consulting firm partners play in implementing purported "high integrity" projects on the ground.

The Role of ESG Metrics in Climate Mitigation: ESG as a Potential Lever for Change

Despite growing skepticism toward ESG metrics, ESG could still play a key role in future climate solutions – if reoriented toward tangible, verifiable impacts. ESG initiatives must evolve beyond performative commitments and focus on:

- Empowering IPLCs
- Reinforcing land tenure security
- Establishing credible incentives that transcend flawed economic models (Benson et al., 2024)

Corporate accountability frameworks, such as those outlined in the Science Based Targets initiative, offer a potential roadmap for ESG improvement. However, without independent third-party verification mechanisms and legally enforceable commitments, ESG will continue to lack credibility in the eyes of investors and policymakers (Harvard Law School Forum on Corporate Governance, 2022).

Beyond Value Chain Mitigation and Scope 3 Challenges

Corporations often highlight improvements in their supply chain management under Scope 3 emissions. However, these measures rarely address governance failures, perverse incentives, and market distortions that drive unsustainable land use in the Global South (Bhattacharya et al., 2020).

Scope 3 emissions – those produced indirectly across a company's supply chain – account for up to 80% of corporate carbon footprints. Yet, voluntary commitments to reduce Scope 3 emissions have largely failed due to inconsistent reporting standards and the absence of financial penalties for non-compliance (Schneider et al., 2024). Without regulatory oversight, these commitments amount to little more than corporate public relations exercises.

When Finance Fails: Lessons From Case Studies in the Global South

Despite significant investments in climate finance, several high-profile carbon offset projects in the Global South have faced serious allegations of human rights abuses, governance failures, and inadequate community engagement. These cases underscore how financial mechanisms, when misaligned with local realities and justice principles, can exacerbate the very issues they aim to mitigate.

Kenya – Kasigau Corridor REDD+ Project

The Kasigau Corridor REDD+ project in Kenya, managed by Wildlife Works, has been self-promoted as a model for market-based conservation. However, a 2023 joint investigation by the Centre for Research on Multinational Corporations (SOMO) and the Kenya Human Rights Commission (KHRC) documented a disturbing pattern of sexual harassment, assault, and coercion spanning over a decade

(Jenkins, 2024; KHRC & SOMO, 2023). Senior male managers allegedly demanded sex in exchange for job retention and denied due process to female employees who came forward. The investigation revealed that the abuse was not isolated but rather systemic, affecting both employees and women in the surrounding communities (FairPlanet, 2023; KHRC & SOMO, 2023).

These abuses raise critical questions about the oversight and ethical governance of internationally funded climate projects. While they could be dismissed as individual incidents, the evidence points to deeper structural failings in the social feasibility of NbS and REDD+ projects run by international intermediaries in remote landscapes of the Global South (Centre for Research on Multinational Corporations (SOMO), 2024).

This case underscores the chapter's central argument that climate finance, when funneled through intermediaries without strong local governance and accountability, risks not only inefficiency but also direct harm to the communities it purports to support. It highlights the urgent need for rights-based, IPLC-led approaches as a prerequisite for legitimate and effective climate mitigation. Fairly negotiated land-use plans are needed if genuine FPIC is to be achieved.

In sum, the Kasigau experience demonstrates that without embedding robust local safeguards and genuine community participation into climate finance mechanisms, even well-intentioned projects can perpetuate confusion, harm, and undermine credibility of both climate and social outcomes. This reinforces the chapter's core thesis: Legitimacy, rights, and governance must be foundational – not optional – for effective mitigation finance, as well as to be marketed as a model for sustainability (United Nations, 2007).

Cambodia – Southern Cardamom REDD+ Project

In Cambodia, the Southern Cardamom REDD+ project, supported by Wildlife Alliance and Wildlife Works Carbon, has faced widespread condemnation for violating Indigenous rights. A 2024 Human Rights Watch report found that the project was implemented without the FPIC of the Indigenous Chong people. The investigation documented forced evictions, arrests for traditional resin tapping, and other repressive tactics that undermined Indigenous livelihoods (Human Rights Watch, 2024). These findings suggest that some REDD+ projects are not just flawed in implementation – they may directly contribute to human rights abuses. The full report is accessible at hrw.org, with a summary coverage at Grist and Flynn (2025).

The Cambodia case powerfully illustrates how climate finance mechanisms that bypass FPIC and local rights can reproduce the very injustices they seek to remedy, reinforcing the chapter's thesis that social legitimacy and community participation are non-negotiable for ethical and sustainable climate action.

Democratic Republic of Congo – Mai-Ndombe REDD+ Project

The Mai-Ndombe REDD+ project in the DRC, also associated with Wildlife Works, has long been controversial. While touted as one of Africa's largest REDD+

pilot programs, critics have pointed to a lack of transparency in benefit-sharing and insufficient engagement with local communities. In response to public scrutiny and third-party rating agency reviews, Wildlife Works issued a formal rebuttal defending the project's methodology and social safeguards. However, questions remain about whether its design meets the social legitimacy standards required for long-term success (Freund, 2024). Their defense of the project can be read on their website: wildlifeworks.com.

These case studies reveal that climate finance – when disconnected from social license, participatory governance, and Indigenous rights – can actively cause harm. REDD+ and similar models must be reimagined not only in financial structure but also in ethical grounding. Only with robust safeguards and IPLC-led governance can such initiatives hope to achieve meaningful climate and social outcomes.

The persistent controversy surrounding the Mai-Ndombe project exemplifies the broader pattern: Without transparency, meaningful community engagement, and secure tenure, even flagship climate finance initiatives are likely to fall short of both environmental and social goals. This reinforces the chapter's argument that reform must begin with governance and rights, not just funding flows. While Mai-Ndombe's notable Coco Pembe advised (Pembe, 2024) that, in fact, a very large transfer of funds had been made to local government by Wildlife Works, as of late 2024 these funds had not been distributed to local communities, reinforcing the perception of governance and corruption weaknesses discussed in Chapter 4.

Addressing Perverse Incentives

A sustainable climate mitigation strategy cannot succeed without directly confronting the perverse incentives that drive environmental degradation, land dispossession, and carbon-intensive development. These incentives – often embedded in laws, subsidies, and market systems – create powerful distortions that undermine conservation efforts and render climate finance ineffective or even counterproductive.

- **Eliminating Harmful Subsidies**:
  Governments around the world continue to provide billions in subsidies for extractive industries such as fossil fuels, industrial agriculture, and large-scale monoculture plantations. These subsidies not only incentivize deforestation, land conversion, and GHG emissions but also make it more difficult for sustainable alternatives to compete (Prakash, 2021). In regions like the Amazon and parts of Southeast Asia, agricultural subsidies for beef, soy, and palm oil have been directly linked to illegal land clearing and biodiversity loss. Unless climate finance is paired with subsidy reform, mitigation investments will be undermined by contradictory incentives that reward destructive practices.
- **Strengthening Policy Enforcement**:
  Even where legal protections exist – such as bans on illegal logging, mining, or encroachment into protected areas – enforcement is often weak or selectively applied. This is particularly true in frontier areas where state presence is minimal and corruption is widespread (Global Witness, 2021). Climate finance

cannot substitute for political will or legal enforcement. In fact, without robust safeguards and local monitoring systems, the influx of finance may exacerbate rent-seeking behavior and accelerate land grabs. Strengthening enforcement must go hand-in-hand with democratizing oversight and empowering local actors, especially IPLCs, to play active roles in accountability systems.

- **Securing Land Tenure**:
  Tenure insecurity remains one of the most deeply rooted structural barriers to effective land-based climate mitigation. When IPLCs lack formal or customary recognition of their land rights, they are vulnerable to displacement, exploitation, and exclusion from benefits, particularly when large-scale finance enters a region. Conversely, securing land tenure has been repeatedly shown to improve environmental outcomes by reinforcing local stewardship systems, reducing deforestation, and aligning conservation with livelihood protection (Frechette et al., 2018). Climate finance that fails to incorporate land tenure reform as a core pillar is likely to reproduce colonial patterns of exclusion under a new green label.
- **Correcting Market Failures**:
  Current market systems routinely fail to account for the environmental and social costs of land degradation, biodiversity loss, and carbon emissions. These externalities allow companies to profit from ecosystem destruction while passing the true costs onto marginalized communities and future generations. To realign incentives, financial systems must begin internalizing these costs – through mechanisms such as environmental taxation, penalties for illegal land use, and credits for ecosystem services (FAO, 2016). However, such mechanisms must be locally adapted and grounded in equity frameworks; otherwise, they risk creating new burdens for already vulnerable communities.

Ultimately, dismantling perverse incentives is not just a technical exercise; it is also a political and, arguably, moral imperative. The core problem is that climate finance has prioritized market-based carbon solutions for what are fundamentally land use and resource rights issues. As a result, it often sidelines the land and resource rights of local communities, defaulting to commodified carbon approaches instead of holistically addressing questions of tenure and land use. Climate finance must no longer reward the least effective actors or reinforce unsustainable systems. It should become a lever for transformation, supporting policies that reward regeneration, uphold rights, and align economic development with long-term ecological and social resilience.

Sovereign Climate Funds and Public Finance Mechanisms: Promise, Pitfalls, and the Case for Inclusive Governance

Sovereign Climate Funds (SCFs) are gaining traction as public finance tools for large-scale climate mitigation, particularly in forest-rich countries. Framed as an alternative to fragmented project finance, SCFs offer the potential for long-term, sovereign-led funding aligned with national priorities. But their impact will depend entirely on their governance: whether they reinforce elite capture and carbon commodification or shift power to IPLCs through participatory structures and social license.

While some analysts view SCFs as a promising evolution in climate finance – offering states fiscal sovereignty and long-term incentives to preserve high-carbon ecosystems (Climate Funds Update, 2024; OECD, 2024) – others caution that such funds risk entrenching elite control, reinforcing opaque institutions, and exacerbating the very issues they are meant to address (Power Shift Africa, 2025; Standing, 2024).

One of the most prominent proposals is the TFFF, a sovereign-backed initiative promoted by Brazil and other Global South countries. The TFFF aims to provide per-hectare payments to countries preserving forests, funded through multilateral guarantees and green bond instruments, with a projected capitalization of $4 billion annually (Climate Change News, 2025). Supporters such as Charlotte Streck see it as a long-overdue shift toward stable, non-market finance grounded in sovereign planning. But others, including Frédéric Hache, caution that its underlying financial assumptions remain opaque and potentially displace historic loss and damage obligations. As Mongabay's DiGirolamo (2024) notes, the TFFF debate encapsulates the core dilemma: Can sovereign climate finance avoid replicating the same extractive logic it seeks to replace? Why would this version of a market-driven model work?

A cautionary case is Bolivia's proposed $1.2 billion sovereign carbon finance deal, which reversed the country's prior rejection of carbon commodification. The agreement has been sharply criticized for its lack of transparency and potential to consolidate power in the hands of political elites, sidelining IPLCs who have historically resisted externally imposed carbon finance models (Power Shift Africa, 2025). The Bolivia deal underscores the risk that SCFs – if not anchored in participatory governance and legal safeguards – can reproduce the same inequities and extractive tendencies as private carbon markets.

In Central Africa, SCF-style mechanisms are rapidly gaining ground. In the Republic of Congo, the government launched the C-15 Climate Justice Initiative, received over $8 million in carbon offset payments, and is promoting natural capital accounting – framing forests as national climate assets.

The BIODEV2030 initiative integrates biodiversity finance into agriculture, mining, and national planning frameworks. In the DRC, several overlapping initiatives reflect similar trends: The Forest Investment Program has delivered over $65 million since 2011, reportedly avoiding 31 million tons of CO_2; the 30×30 Strategy Phase 2, supported by the Green Climate Fund, aims to protect 30% of national territory by 2030; and the IMF's Resilience and Sustainability Facility is financing public sector reforms and climate integration.

Yet in stark contradiction, the DRC government announced plans in 2022 to auction 30 oil and gas blocks – at least nine overlapping with the Cuvette Centrale peatlands, which hold roughly 30 billion metric tons of carbon (Cannon, 2022; Dargie et al., 2022; WCS, 2024). This juxtaposition lays bare the central tension: SCFs cannot deliver legitimacy or climate impact if they reinforce extractive models cloaked in green rhetoric.

For SCFs to succeed, they must be embedded in transparent, inclusive governance frameworks that generate social license – not just financial flows. That means:

- **Tripartite co-governance platforms**, where IPLCs, governments, and donors jointly define rules, responsibilities, and safeguards through negotiation;
- **Direct access pathways** for IPLC institutions to receive funds without relying on intermediaries;
- **Legal protection of customary tenure systems** to prevent elite capture and ensure land rights are enforceable;
- **Performance-based disbursements**, tied to locally defined indicators of ecological health, livelihood resilience, and tenure security – not just carbon.

Without these foundations, SCFs risk becoming the next iteration of extractive climate finance. With them, however, they could become a cornerstone of a new public finance architecture – one that replaces carbon commodification with justice-based stewardship, negotiated consent, and long-term climate legitimacy.

The Political Economy of Climate Mitigation in the Global South: Carbon Markets, Structural Barriers, and the Need for Negotiated Governance

Carbon markets and offset schemes are frequently promoted as efficient tools for channeling climate finance to the Global South, with advocates claiming they can deliver both emissions reductions and sustainable development benefits (Environmental Finance, 2024; Green Central Banking, 2025). However, a growing body of research and critical analysis demonstrates that these mechanisms have not only failed to deliver meaningful benefits for vulnerable communities but have also deepened structural barriers to equitable and effective climate action (Cannon, 2024; Carbon Market Watch, 2024; Gabbatiss et al., 2023).

Market-Driven Mechanisms and Marginalization

Carbon markets are widely presented as a win-win for climate and development, promising to direct finance to forest protection, low-carbon development, and local livelihoods in the Global South (Environmental Finance, 2024; Wildlife Works, 2024). Yet in practice, the majority of financial flows generated by these mechanisms accrue to project developers, intermediaries, and corporations – primarily based in the Global North – while only a small proportion reaches IPLCs or supports genuine climate action on the ground (Carbon Market Watch, 2024; NewClimate Institute, 2024).

A recent Carbon Market Watch study found that 62% of companies involved in African carbon projects were based in countries with "very high human development," while less than 28% were actually from African countries. This pattern holds true across all types of actors in the voluntary carbon market, including project developers, owners, consultants, and validators, raising serious questions about who really benefits from these schemes (Carbon Market Watch, 2024).

Despite claims that carbon offsetting unlocks finance for the Global South, there is little evidence that the marketed climate and financial benefits actually reach local communities. Numerous reports indicate that the primary beneficiaries are project developers, international traders, and brokers based in the Global North.

There is no robust system to ensure that finance from carbon credit projects is directed to intended beneficiaries, resulting in opaque operations and a lack of transparency about fees, profits, and the actual flow of funds. Nine out of ten intermediaries in voluntary carbon markets do not disclose their fees or profit margins, and project developers and owners are predominantly based in the Global North (NewClimate Institute, 2024).

Land Dispossession and "Carbon Colonialism"

The expansion of carbon offset projects has led to large-scale land acquisitions by foreign companies, often displacing local communities and undermining traditional land rights. A recent report found that communal lands the size of Portugal have been taken over for carbon offsetting schemes across the Global South, with more than half of this land in Africa. In most of the 279 projects surveyed, foreign companies acquired large tracts of land traditionally used for grazing or farming and converted them into monoculture plantations for carbon sequestration. This process, described as a form of "carbon colonialism," prioritizes the interests of corporations over local livelihoods and food security (Hanbury, 2024).

Divergence of Perspectives and Calls for Reform

While some Indigenous Peoples and local community organizations in the Global South advocate for greater inclusion in carbon market design and implementation, seeing potential for these mechanisms to provide much-needed finance for conservation and livelihoods, there is also widespread concern about the risks of exclusion, lack of self-determination, and inequitable benefit sharing (Banarjee, 2024).

Community-based organizations and small businesses from across the Global South have repeatedly called for reforms to ensure that carbon markets deliver real, measurable benefits to those on the frontlines of climate change, rather than enriching external actors (Banarjee, 2024; NewClimate Institute, 2024). Despite these calls, the current structure of carbon markets continues to reinforce a North-South divide, with the Global South bearing the social and ecological costs of offsetting schemes while the Global North captures the majority of financial returns and emissions accounting benefits (Carbon Market Watch, 2024; NewClimate Institute, 2024).

Toward Transformative Land Use and Governance

Given these persistent barriers, there is an emerging recognition that climate mitigation in the Global South requires a shift from market-based mechanisms to approaches that elevate land use as a central component of climate and development strategy. INRM is increasingly recognized as a framework that can address these challenges by centering local agency, ecological integrity, and negotiated governance. INRM emphasizes community-led stewardship of land, water, and ecosystems and seeks to harmonize conservation, production, and social equity at the landscape level.

Central to this approach is the concept of Negotiated Governance Platforms (NGPs), which will be developed in Chapters 6 and 7. NGPs provide a mechanism for inclusive, participatory decision-making that brings together diverse stakeholders to negotiate land-use priorities, rights, and responsibilities. Through NGPs, INRM offers the potential to overcome the structural barriers of the current paradigm, ensuring that climate action is grounded in local realities and priorities and that benefits are equitably shared.

The Great Green Wall: A Cautionary Tale of Ambition Without Architecture

Few initiatives capture the tensions at the heart of climate finance better than the Great Green Wall (GGW) – a high-profile, multibillion-dollar effort that embodies both the ambition of large-scale land restoration and the institutional failures that plague implementation in the Global South. Like many climate finance mechanisms examined in this chapter, the GGW has struggled not for lack of vision, but for lack of governance, equity, and social license.

The GGW remains one of the most ambitious land restoration efforts in the world – promising to restore 100 million hectares across the Sahel, sequester 250 million tons of carbon, and create 10 million rural jobs. But as of 2025, progress is still lagging far behind targets. Roughly 10% of the projected area has been covered – up from just 4% in 2020 – but survival rates of planted trees remain low, and the gains in livelihoods and ecosystem restoration are scattered and uneven (Pan-African Agency of the Great Green Wall (GGW Agency), 2025; Zhu et al., 2025).

The GGW exemplifies the core dilemma of climate finance in the Global South: high-profile pledges without the institutional scaffolding to deliver equitable, sustainable outcomes. Despite billions committed to the Great Green Wall, progress has been slowed by administrative bottlenecks, coordination challenges, and uneven funding flows among participating countries. The UNCCD (2023) notes that while commitments are substantial, comparable tracking of disbursements and on-the-ground delivery remains limited, and the proportion of funds reaching restoration activities cannot be reliably quantified.

While some localized successes in agroforestry and drought-resilient planting have been documented, these tend to emerge where communities have been engaged directly in planning and monitoring. But such cases remain the exception rather than the rule. Many local communities report limited say in decision-making and uneven benefit sharing – echoing broader patterns seen across international conservation finance efforts (African Forest Landscape Restoration Initiative, 2025; Environmental Defender's Office, 2019; UNEP, 2025).

More fundamentally, the GGW exposes the limits of a model that scales ambition without building governance. The initiative has made incremental progress in ecological adaptation – such as deploying drought-tolerant species – but still lacks stable, long-term financing tied to community-defined outcomes. And without addressing core issues like land rights, equitable benefit distribution, and IPLC participation, the project risks becoming another example of institutional drift: a well-intentioned initiative bogged down by bureaucracy, vulnerable to elite capture, and disconnected from the people and ecosystems it was meant to serve.

The lesson is not to abandon such landscape-scale programs but to reimagine them: Root them in negotiated governance, ensure direct IPLC participation and funding access, and align incentives with stewardship rather than symbolic targets. The GGW's future will depend not on planting more trees but on building the political and institutional soil where real restoration can take root.

Moving Forward: Aligning Climate Mitigation With Justice and Governance

To transition toward meaningful and lasting climate action, reforms must go beyond financial ambition and address the structural drivers of failure. The following pillars represent not just technical priorities but also foundational shifts necessary to bring justice, legitimacy, and effectiveness into climate finance systems.

Corporate accountability must move beyond performative ESG commitments and be grounded in verifiable, ground-level impacts. Current ESG frameworks too often reward disclosure over outcomes and allow corporations to claim environmental virtue while supporting extractive supply chains. What is needed instead are metrics tied to land-use practices, biodiversity integrity, and social safeguards – benchmarks that can be independently verified and aligned with IPLC-defined priorities. Without enforceable standards, ESG remains a branding exercise. Building on Schneider et al. (2024), integrity in ESG must be rooted in substance, not surface compliance.

Regulatory enforcement is essential for translating policy into practice. Many land-based climate initiatives have failed not because of weak design but because enforcement institutions are either under-resourced or politically compromised. Land grabbing, illegal logging, and resource extraction flourish in governance vacuums, particularly in contested forest regions. Closing these loopholes requires not only formal legal tools but also community monitoring and grievance mechanisms that empower local actors to uphold environmental laws. Human Rights Watch (2021) highlights how impunity for environmental crimes in the Amazon persists despite formal protections – underscoring the need for credible enforcement as a precondition for climate integrity.

Empowering IPLCs must become a foundational goal, not a co-benefit or afterthought. Indigenous and local communities steward a significant share of the world's remaining biodiversity and carbon-rich ecosystems, often under customary tenure systems that lack formal recognition. Yet too often, they are treated as implementation partners, not governance actors. Real empowerment means embedding IPLC knowledge systems and decision-making authority into the design, governance, and evaluation of climate mitigation efforts. It also means respecting FPIC, legally securing collective tenure, and financing IPLC institutions directly. As the FAO (2016) has shown, the sustainability of land-use initiatives is directly correlated with the degree of community ownership and governance.

Rethinking finance involves shifting the emphasis from quantity to quality. Scaling finance is important, but where that finance goes – and how it's governed – matters far more. Funding must support governance reform, tenure security, and negotiated platforms where IPLCs, governments, and other stakeholders define goals together. Without such foundational investments, large-scale funding will continue to be absorbed by intermediaries or channeled into inappropriate STAM strategies that produce short-term results at long-term social and ecological cost. Bhattacharya et al. (2020) stress that effective climate finance must address power imbalances, not just resource gaps.

The prevailing model of climate finance and mitigation commitments must be overhauled to prioritize justice, accountability, and sustainable land governance. Without addressing these systemic issues – corporate evasion, legal loopholes, exclusion of IPLCs, and misplaced financial incentives – climate mitigation efforts risk reinforcing the very challenges they aim to solve.

Rethinking Climate Finance Through Social License and Governance Reform

The persistent failure of climate finance to deliver equitable, effective outcomes in the Global South reflects deeper structural flaws, not just funding shortfalls. While it is tempting to view the crisis as a matter of insufficient money, this framing obscures the real barriers: weak governance, flawed delivery models, and the failure to generate social legitimacy. Standing (2024) argues that in the context of biodiversity finance, fixating on headline funding gaps can be "dangerous nonsense" when the core issues are institutional and political.

As this chapter has shown, the current architecture too often reinforces extractive models, bypasses IPLCs, and clings to flawed market-based instruments. Rather than centering carbon metrics and hypothetical co-benefit sharing through non-transparent partnerships, climate finance must prioritize sustainable land use and create tangible, direct incentives for ecosystem stewardship.

Rethinking climate finance means moving beyond pledges and price signals toward negotiated governance, tenure security, and performance-based mechanisms that generate legitimacy on the ground. Only through social license, institutional reform, and context-appropriate STAM can climate finance become a tool for transformation rather than a perpetuation of inequity.

To truly transform climate finance in the Global South, a shift is needed away from market-driven models toward finance structures that prioritize local governance, Indigenous rights, and participatory approaches. This transformation requires moving beyond carbon markets that focus narrowly on carbon offsets and instead supporting community-led climate mitigation strategies.

Chapters 6 and 7 explore why and how integrated land-use systems – specifically INRM, NGPs, and IPLC empowerment – must drive a fundamental shift in climate finance. Direct access to finance for communities, with funds tied to socially legitimate, technically feasible, and context-specific mitigation strategies, is essential. This approach ensures that climate finance supports sustainable, locally driven solutions, rather than reinforcing the power structures that have historically marginalized Indigenous and local communities through top-down planning and international agendas that prioritize carbon accounting over common sense and local realities.

Note

1 Quote from Walter Prahl: https://www.linkedin.com/in/walter-prahl-b58ab5110/.

References

ActionAid International. (2023). Annual Report 2023 – Hope & Humanity. ActionAid International. https://actionaid.org/publications/2024/annual-report-2023-hope-humanity

Adesina, A. (2025, April 6). 'Land grabs' in Africa replaced by 'carbon grabs', says bank chief. *Financial Times*. Retrieved from https://www.ft.com/content/0468305c-00b0-47fe-bfef-4a3f8d0b3065

African Climate Wire. (2025, May 16). Africa Great Green Wall Initiative Seeks to Overcome Policy and Funding Gaps. https://africanclimatewire.org/2025/05/africa-great-green-wall-initiative-seeks-to-overcome-policy-and-funding-gaps/

African Forest Landscape Restoration Initiative (AFR100). (2025). *Monitoring: How to Promote and Measure Landscape Restoration*. https://afr100.org/monitoring

Alayza, N., & Larsen, G. (2025, February 20). How to reach $300 billion – and the full $1.3 trillion – under the new climate finance goal. *World Resources Institute*. https://www.wri.org/insights/ncqg-climate-finance-goals-explained

Amel-Zadeh, A., & Serafeim, G. (2018). *Why and how investors use ESG information: Evidence from a global survey*. *Financial Analysts Journal*, 74(3), 87–103. https://www.tandfonline.com/doi/full/10.2469/faj.v74.n3.2

Anozie, C. (2024). Left Behind: Funding Climate Action in the Global South. *Texas A&M Law Review*, 11(2), 333–372.

Atal, R., & Sylvan, D. (2025, September). Integrity, Equivalence, and Imperfect Carbon Offsets. *Institute for Policy Integrity*. Retrieved from https://policyintegrity.org/files/publications/Integrity_Equivalence_and_Imperfect_Carbon_Offsets.pdf

Bain & Company. (2025). *Oil and Gas Executive Perspectives: Energy Transition Timelines Extend to 2070*. April 2025 Report.

Banerjee, A. (2024, May 6). S3|EP20 - Carbon credits & conservation: How Wildlife Works is saving forests and communities | Mike Korchinsky. *The Think Wildlife Podcast*. https://anishbanerjee.substack.com/p/s3ep20-carbon-credits-and-conservation

Belfer Center, Harvard Kennedy School. (2024). Article 6 at COP29: The challenge for carbon markets is demand not supply. https://www.belfercenter.org/research-analysis/article-6-cop29-challenge-carbon-markets-demand-not-supply

Benson, S., Farrelly, A., Watson, E., Kazanecki, H., Massei, M., von Preussen, A., Streck, C., & Trouwloon, D. (2024). *Above and beyond: An SBTi Report on the Design and Implementation of BVCM*. Science Based Target Initiative.

Bhattacharya, A., Calland, R., Averchenkova, A., Gonzalez, L., Martinez-Diaz, L., & Van Rooij, J. (2020). *Delivering on the $100 Billion Climate Finance Commitment and Transforming Climate Finance*. Independent Expert Group on Global Finance.

Bhuyan, L. M. (2025). Refocusing conservation funding on trust & community leadership after USAID freeze (commentary). *Mongabay*. https://news.mongabay.com/2025/04/refocusing-conservation-funding-on-trust-community-leadership-after-usaid-freeze-commentary/

BioCarbon Partners. (2025). Zambia's Luangwa Community forests project. https://bcp.earth/news-and-press/zambias-luangwa-community-forests-project-wins-worlds-best-carbon-offset-project/

Boell Foundation. (2023). COP 28: Why global carbon pricing and trading platforms are false solutions. https://www.boell.de/en/2023/11/17/article-6-cop-28-why-global-carbon-pricing-and-trading-platforms-are-false-solutions

Boffo, R., and R. Patalano (2020), "ESG Investing: Practices, Progress and Challenges", OECD Paris, www.oecd.org/finance/ESG-Investing-Practices-Progress-and-Challenges.pdf

Boudreaux, Karol. (2010). *Community Conservation in Namibia: Devolution as a Tool for the Legal Empowerment of the Poor*. International Development Law Organization (IDLO) Working Paper Series. Mercatus Center, George Mason University.

Bretton Woods Project. (2024). COP29 sees MDBs climate finance take centre stage, as civil society brands new climate finance goal a "betrayal". Retrieved from https://www.brettonwoodsproject.org/2024/12/cop29-sees-mdbs-climate-finance-take-centre-stage-as-civil-society-brands-new-climate-finance-goal-betrayal/

Brookings. (2024). Identifying barriers to advancing climate equity through U.S. financial markets. https://www.brookings.edu/articles/identifying-barriers-to-advancing-climate-equity-through-u-s-financial-markets/

Buchner, B., Naran, B., & Oliver, P. (2023). Global Landscape of Climate Finance 2023. *Climate Policy Initiative*. https://www.climatepolicyinitiative.org/publication/global-landscape-of-climate-finance-2023/

Callahan, C. W., & Mankin, J. S. (2025). Carbon majors and the scientific case for climate liability. *Nature*, 640, 893–901. https://doi.org/10.1038/s41586-025-08751-3

Cannon, J. (2022, July 28). In Congo, a carbon sink like no other risks being carved up for oil. *Mongabay*. https://news.mongabay.com/2022/07/in-congo-a-carbon-sink-like-no-other-risks-being-carved-up-for-oil/

Cannon, J. (2022a, November 23). Despite pledges, obstacles stifle community climate and conservation funding. *Mongabay*. https://news.mongabay.com/2022/11/despite-pledges-obstacles-stifle-community-climate-and-conservation-funding/

Cannon, J. (2024, December 23). The state of carbon markets in 2024. *Mongabay*. https://news.mongabay.com/2024/12/the-state-of-carbon-markets-in-2024/

CarbonCredits.com. (2025). What is the voluntary carbon market? https://carboncredits.com/what-is-the-voluntary-carbon-market/

Carbon Market Watch. (2024, July 29). Due South: Voluntary carbon market's North-South inequality under the spotlight. https://carbonmarketwatch.org/2024/07/29/due-south-vcms-north-south-inequality-under-the-spotlight/. PDF report: https://carbonmarketwatch.org/wp-content/uploads/2024/07/Final_Due-South_Geographic-disparity-of-project-actors-in-the-voluntary-carbon-market.pdf

Carbon Pricing Leadership Coalition. (2019). Report of the High-Level Commission on Carbon Pricing and Competitiveness. World Bank. http://hdl.handle.net/10986/32419

Centre for Research on Multinational Corporations (SOMO). (2024). *Beyond the Project: Structural Barriers to Social Feasibility in Nature-based Solutions and REDD+ Initiatives*. SOMO.

CIFOR-ICRAF. (2021). Seeds of adaptation: Women farmers, local seed systems, and climate change. *Gender Climate Brief*. https://www.cifor-icraf.org/publications/pdf_files/brief/5896-GenderClimateBrief.pdf

Climate Change News. (2025, June 5). Brazil's new funding plan can help bring rainforests back – if done right. https://www.climatechangenews.com/2025/06/05/brazils-new-funding-plan-can-help-bring-rainforests-back-from-the-brink/

Climate Funds Update. (2024). *Global Climate Finance Architecture: Annual Update 2024*. Climate Funds Update. ODI & Heinrich Böll Stiftung.

Climate Impact X. (2022, November 1). CIX completes first-of-its-kind portfolio auction of voluntary nature-based carbon credits with leading global companies. https://climateimpactx.com/press-release/cix-completes-first-of-its-kind-portfolio-auction-of-voluntary-nature-based-carbon-credits-with-leading-global-companies/

Climate Investment Funds. (2024, March 19). Climate investment funds earmarks $40 million for Indigenous-led climate action in seven countries. https://www.cif.org/news/climate-investment-funds-earmarks-40-million-indigenous-led-climate-action-seven-countries-0

Climate Policy Initiative. (2022, November 10). Gender-smart climate finance: Three things you need to know ahead of COP27. https://www.climatepolicyinitiative.org/gender-smart-climate-finance-three-things-you-need-to-know-ahead-of-cop27/

Convergence & Climate Policy Initiative. (2024, September 1). Blended finance and the gender-energy nexus: A stocktaking report. https://www.climatepolicyinitiative.org/publication/blended-finance-and-the-gender-energy-nexus-a-stocktaking-report/

Council on Energy, Environment and Water (CEEW). (2025). Accelerating Investments for Nature-based Solutions in the Global South: A Unified Framework for Mapping and Estimating Benefits. ENSURE – Effective Nature-based Solutions Utilisation and Resource Evaluation Framework. CEEW, New Delhi. https://www.ceew.in/sites/default/files/how-can-investing-in-nature-based-solutions-for-climate-change-enhance-global-climate-action.pdf

Dale, G., & Holmes, C. (Eds.). (2023). *Karl Polanyi's Political and Economic Thought: A Critical Guide*. Cambridge University Press.

Dargie, G. C., Crezee, B., Biddulph, G., et al. (2022). Mapping peat thickness and carbon stocks of the central Congo Basin. *Nature Geoscience*, 15, 711–718.

Development Initiatives. (2024, September 2). From gender-relevant to gender-transformative climate finance: Rwanda case study. https://devinit.org/resources/gender-relevant-to-transformative-climate-finance-rwanda/

DiGirolamo, M. (2024, December 16). What's the TFFF? A forest finance tool 'like no other' shows potential. *Mongabay*. https://news.mongabay.com/2024/12/whats-the-tfff-a-forest-finance-tool-like-no-other-shows-promise-pitfalls/

Earth.Org. (2025). USAID funding freeze deals a major blow to climate finance. https://earth.org/tag/us-agency-for-international-development-usaid/

Ecosystem Marketplace. (2024). *2024 State of the Voluntary Carbon Market (SOVCM)*. Washington, DC: Forest Trends.

Environmental Defender's Office. (2019). *PNG Conservation Projects Face Allegations of Land-Grabbing*.

Environmental Finance. (2024). The Global South can't beat climate change without voluntary carbon markets. https://www.environmental-finance.com/content/analysis/the-global-south-cant-beat-climate-change-without-voluntary-carbon-markets.html

Eriksen, S., et al. (2021). Adaptation interventions and their effect on vulnerability in developing countries: Help, hindrance or irrelevance? *World Development*, 141, 105383. https://doi.org/10.1016/j.worlddev.2020.105383

Euronews Green. (2023). Climate funds for Indigenous Peoples "evaporate" before reaching them. https://www.euronews.com/green/2023/12/05/climate-funds-for-indigenous-peoples-evaporate-before-reaching-them-report-reveals

FairPlanet. (2023). The sexual abuse case rocking the carbon offsetting industry. https://www.fairplanet.org/story/the-sexual-abuse-case-rocking-the-carbon-offsetting-industry/

FAO & Global Land Alliance. (2025). *Collective Tenure Rights and Climate Action in Sub-Saharan Africa*. Rome and Washington, DC: FAO & Global Land Alliance.

Flynn, G. (2025, March 11). Indigenous community calls out Cambodian REDD+ project as tensions simmer in the Cardamoms. *Mongabay.* https://news.mongabay.com/2025/03/indigenous-community-calls-out-cambodian-redd-project-as-tensions-simmer-in-the-cardamoms/

Food and Agriculture Organization of the United Nations (FAO). (2012). *Voluntary Guidelines on the Responsible Governance of Tenure of Land, Fisheries and Forests in the Context of National Food Security*. Rome: FAO. https://www.fao.org/tenure/voluntary-guidelines/en/ and https://www.fao.org/3/i2801e/i2801e.pdf

Food and Agriculture Organization of the United Nations (FAO). (2016). *State of Food and Agriculture 2016: Climate Change, Agriculture and Food Security*. Rome: FAO. https://www.fao.org/3/i6030e/i6030e.pdf

Food and Agriculture Organization of the United Nations (FAO). (2023). *The Status of Women in Agrifood Systems*. Rome: FAO.

Frechette, A., Ginsburg, C., Walker, W., Gorelik, S., Keene, S., Meyer, C., Reytar, K., & Veit, P. (2018). A global baseline of carbon storage in collective lands: Indigenous and local community contributions to climate change mitigation. *Rights and Resources Initiative.* https://rightsandresources.org/wp-content/uploads/2018/09/A-Global-Baseline_RRI_Sept-2018.pdf

Freund, J. (2024, April). Mai Ndombe REDD+ Project: Wildlife Works Response to Inaccurate Ratings Agency Review. Response to Renoster's "Pro Review" of the Mai Ndombe REDD+ Project. *Wildlife Works*. https://www.wildlifeworks.com/post/raising-the-baar-innovative-research-in-forest-conservation-helps-guide-the-voluntary-carbon-market

Gabbatiss, J., Dunne, D., Chandrasekhar, A., Dwyer, O., Lempriere, M., Quiroz, Y., Tandon, A., & Viglione, G. (2023, September 24). In-depth Q&A: Can 'carbon offsets' help to tackle climate change? *Carbon Brief. Design* by J. Goodman, T. Pearson, & T. Prater. Available at: https://interactive.carbonbrief.org/carbon-offsets-2023/index.html

Garnett, S. T., Burgess, N. D., Fa, J. E., Fernández-Llamazares, Á., Molnár, Z., Robinson, C. J., & Leiper, I. (2018). A spatial overview of the global distribution of indigenous and local community lands. *World Development*, 104, 112–122.

Global Center on Adaptation (GCA). (2024). *GCA Leads Locally Led Adaptation Efforts in Senegal River Valley.* Available at: https://gca.org/gca-leads-locally-led-adaptation-efforts-in-senegal-river-valley/

Global Environment Facility Small Grants Programme (GEF SGP). (2024). *Results Report 2023–2024*. United Nations Development Programme. https://www.thegef.org/newsroom/publications/gef-small-grants-programme-results-report-2023-2024

Global Witness. (2021). *Defending Tomorrow: The Climate Crisis and Threats against Land and Environmental Defenders*. London: Global Witness.

Goulder, L. H., & Hafstead, M. A. C. (2022). Why the US should establish a carbon price either through reconciliation or other legislation. *Brookings Institution*. https://www.brookings.edu/articles/why-the-us-should-establish-a-carbon-price-either-through-reconciliation-or-other-legislation/

Green Central Banking. (2025, April 15). Carbon markets and unlocking green finance for the global south. Retrieved from https://greencentralbanking.com/2025/04/16/carbon-markets-unlocking-green-finance/

Grown, C., Heckwolf, A., & Soubeyran, E. (2024). Why gender equality must be part of global climate finance reform initiatives. *Brookings Institution*. https://www.brookings.edu/articles/why-gender-equality-must-be-part-of-global-climate-finance-reform-initiatives/

Hamerkop. (2023). The ever-expanding world of carbon certification standards. https://www.hamerkop.co/blog/the-ever-expanding-world-of-carbon-certification-standards

Hanbury, S. (2024). Carbon credit land grab dispossesses Global South communities: Report. https://news.mongabay.com/short-article/carbon-credit-land-grab-dispossesses-global-south-communities-report/

Hann, Chris. (2021). One hundred years of substantivist economic anthropology. *Max Planck Institute for Social Anthropology Working Paper No. 220*. https://www.scribd.com/document/871637423/One-Hundred-Years-of-Substantivist-Economic-Anthropology

Harvard Law School Forum on Corporate Governance. (2022, June 17). ESG global study 2022. https://corpgov.law.harvard.edu/2022/06/17/esg-global-study-2022/

Heeb, G. (2025, January 2). Big Banks Flee Climate Coalition Formed to Reduce Carbon Emissions: JPMorgan Chase, the last major U.S. bank in the coalition, could also withdraw. *The Wall Street Journal*. https://www.wsj.com/finance/banking/big-banks-flee-climate-coalition-formed-to-reduce-carbon-emissions-72923e2f

High-Level Commission on Carbon Prices. (2017). *Report of the High-Level Commission on Carbon Prices*. Washington, DC: World Bank. https://openknowledge.worldbank.org/entities/publication/f1439afb-82a8-5a92-b95f-0f224da85169

Hodel, L., le Polain de Waroux, Y., & Garrett, R.D. (2024). Characterizing culture's influence in land systems. *Nat Sustain* 7, 973–982. https://doi.org/10.1038/s41893-024-01381-z

Human Rights Watch. (2021). Rainforest Mafias: How violence and impunity fuel deforestation in Brazil's Amazon. https://www.hrw.org/report/2019/09/17/rainforest-mafias/how-violence-and-impunity-fuel-deforestation-brazils-amazon

Institute for Environment and Development (IIED). (2024, June). *Finance for Climate Adaptation in the Most Vulnerable Places. . . Time for a Major Rethink*. IIED Insights.

Intergovernmental Panel on Climate Change (IPCC). (2019). Climate change and land: An IPCC special report on climate change, desertification, land degradation, sustainable land management, food security, and greenhouse gas fluxes in terrestrial ecosystems. https://www.ipcc.ch/srccl/

Intergovernmental Panel on Climate Change (IPCC). (2022). *Climate Change 2022: Mitigation of Climate Change. Contribution of Working Group III to the Sixth Assessment Report*. Cambridge University Press.

Interlaken Group & Rights and Resources Initiative (RRI). (2023 May 25). *Principles of Community Monitoring: A Path for Companies and Investors to Strengthen Human Rights and Environmental Due Diligence and Support Community Tenure in Land-Based Sectors*. Washington, D.C.: Rights and Resources Initiative. https://www.interlakengroup.org/community-monitoring

International Institute for Environment and Development (IIED). (2024). *Finance for climate adaptation in the most vulnerable places . . . Time for a major rethink*. IIED Insights, June 2024.

International Swaps and Derivatives Association (ISDA). (2023). Navigating the risks of greenwashing in the voluntary carbon market. https://www.isda.org/2024/04/17/navigating-the-risks-of-greenwashing-in-the-voluntary-carbon-market/

IUCN. (2024). *IUCN Work with Global and Regional Forest and Farm Producer Organizations: Regional and Global Highlights in 2024*. Forest and Farm Facility.

Jenkins, M. (2024). The impacts of corruption on women and their role in preventing corruption. Reviewed by M. Kirya. Transparency International & U4 Anti-Corruption Resource Centre. Retrieved from https://www.u4.no/publications/the-impacts-of-corruption-on-women-and-their-role-in-preventing-corruption

Kenya Human Rights Commission (KHRC) & SOMO. (2023). *Offsetting Human Rights: Sexual abuse and harassment at the Kasigau Corridor REDD+ Project in Kenya*. KHRC & SOMO.

LandLinks. (2024). *Gender and Property Rights in the Sahel: Implications for Climate Adaptation*. USAID LandLinks.

Lang, C. (2024, November 12). COP29 approval of Article 6.4 is "opening the floodgates for a global carbon market that will have devastating impacts on communities in the Global South." *REDD-Monitor*. https://reddmonitor.substack.com/p/cop29-approval-of-article-64-is-opening

NACSO. (2024). The 1996 Nature conservation amendment act established conservancies as legal entities, enabling communities to manage wildlife and tourism revenues directly. https://www.nacso.org.na/what-we-do/conservancies

Nature Sustainability. (2024). Ecocultural drivers of land management: recognizing kinship, risk, spirituality, and trauma in socio-ecological decision-making. *Nature Sustainability*, 7(3), 205–212.

NewClimate Institute. (2024). The promise of voluntary carbon markets unlocking finance for the Global South may be a myth. https://newclimate.org/news/the-promise-of-voluntary-carbon-markets-unlocking-finance-for-the-global-south-may-be-a-myth

Oliver Wyman. (2023). Applying a gender lens to climate investing. *Oliver Wyman Forum*. https://www.oliverwyman.com/our-culture/society/social-impact/applying-a-gender-lens-to-climate-investing.html

Organisation for Economic Co-operation and Development (OECD). (2024). Climate finance and the USD 100 billion goal. *OECD*. https://www.oecd.org/en/topics/sub-issues/climate-finance-and-the-usd-100-billion-goal.html

Pan-African Agency of the Great Green Wall (GGW Agency). (2025). *Great Green Wall Implementation Status and Way Ahead to 2030*. Bonn: UNCCD Secretariat.

Pembe, C. (2024). Personal Communication. Kinshasa.

Perkins, N. (2024, October 16). The criticism of ESG: why is it becoming controversial? *University of the Built Environment*. Retrieved from https://www.ube.ac.uk/whats-happening/articles/criticism-of-esg/

Polanyi, Karl. 1944. *The Great Transformation*. New York: Farrar & Rinehart/Beacon Press.

Porras, I. and Chacón-Cascante, A. (2018). Costa Rica's payments for ecosystem services programme. https://www.iied.org/g04272

Power Shift Africa. (2025). Carbon markets are a failed model of climate finance that prioritises profits over people and delays over action. As cited in: Lang, C. (2025, July 4). REDD Monitor Substack. Retrieved from https://reddmonitor.substack.com/p/carbon-markets-are-a-failed-model

Prakash, A. (2021). *Repurposing Perverse Incentives for Land Restoration. UNCCD Global Land Outlook Working Paper*. Bonn: United Nations Convention to Combat Desertification (UNCCD).

Rainforest Foundation UK. (2024). A renewed focus on direct financing at international climate summits. https://rainforestfoundation.org/a-renewed-focus-on-direct-financing-at-international-climate-summits/

Riehl B., Zerriffi H., Naidoo R. (2015) Effects of community-based natural resource management on household welfare in Namibia. *PLoS ONE*, 10(5): e0125531. https://journals.plos.org/plosone/article?id=10.1371/journal.pone.0125531

Rights and Resources Initiative. (2020, August). The Opportunity Framework 2020: Identifying Opportunities to Invest in Securing Collective Tenure Rights in the Forest Areas of Low- and Middle-Income Countries. Retrieved from https://rightsandresources.org/wp-content/uploads/Opp-Framework-v2021.pdf

Rights and Resources Initiative (RRI). (2022). *Annual Report 2022*. Washington, DC: Rights and Resources Initiative. https://rightsandresources.org/publication/annual-report-2022/

Schneider, L., Füssler, J., Oberpriller, Q., & Spalding-Fecher, R. (2024). The ICVCM approval of three REDD methodologies presents risks to the integrity of the initiative. *Oeko-Institut*. https://www.oeko.de/blog/the-icvcm-approval-of-three-redd-methodologies-presents-risks-to-the-integrity-of-the-initiative/

SEI. (2020). Why is it so hard to spend climate finance? https://www.sei.org/perspectives/why-is-it-so-hard-to-spend-climate-finance/

SIPRI. (2025). Closing climate security gaps requires more reliable data on climate finance. https://www.sipri.org/commentary/blog/2025/closing-climate-security-gaps-requires-more-reliable-data-climate-finance

Solidaridad. (2022). Climate finance: The need for effective and locally-led climate finance to build farmer climate resilience. https://www.solidaridadnetwork.org/wp-content/uploads/2022/11/2022-Climate-Finance-Paper-Solidaridad-COP27-final.pdf

Standing, A. (2024). Why the $700 billion funding gap for biodiversity is dangerous nonsense: Implications for the oceans and small-scale fisheries. *Coalition for Fair Fisheries Arrangements*. https://www.thecornerhouse.org.uk/resource/why-700-billion-funding-gap-biodiversity-dangerous-nonsense

Stankova, T. (2025, April 8). African Development Bank President criticizes low carbon credit prices and foreign exploitation. Carbon Herald. https://carbonherald.com/african-development-bank-president-criticizes-low-carbon-credit-prices-and-foreign-exploitation/

Sylvera. (2023, March 24). Carbon markets 2023 pulse check. https://www.sylvera.com/blog/carbon-markets-2023-check-in

Timperley, J. (2021, July 27). The truth behind corporate climate pledges. *The Guardian*. https://www.theguardian.com/environment/2021/jul/27/the-truth-behind-corporate-climate-pledges-jocelyn-timperley

United Nations. (2007). *United Nations Declaration on the Rights of Indigenous Peoples*. Resolution adopted by the General Assembly, 13 September 2007, A/RES/61/295. New York: United Nations.

United Nations Convention to Combat Desertification (UNCCD). (2023). Independent Review of the Great Green Wall Accelerator: Final Report, February 2023. Bonn, Germany: UNCCD Secretariat. Retrieved from https://www.unccd.int/sites/default/files/2023-02/Independent%20Review%20of%20the%20Great%20Green%20Wall%20Accelerator%20Final%20Report.pdf

United Nations Development Programme (UNDP). (2024). *Training module 5: Gender and climate finance*. https://www.undp.org/sites/g/files/zskgke326/files/publications/Gender_Climate_Change_Training%20Module%205%20Finance.pdf

United Nations Environment Programme (UNEP). (2025). *State of Finance for Nature*. UNEP Climate Finance Unit.

UNFCCC. (2023). Gender Action Plan: Enhanced Lima work programme on gender and its gender action plan: Progress and results (FCCC/SBI/2023/8).

United Nations Framework Convention on Climate Change (UNFCCC). (2024). New collective quantified goal on climate finance. *UNFCCC*. https://unfccc.int/NCQG

UN Women. (2024). *Gender, Climate Finance, and Equitable Adaptation: Lessons from East Africa*. New York: UN Women Policy Brief Series, Issue 18.

Wildlife Conservation Society (WCS). (2024). WCS High-Integrity Peatlands Cross-Regional Initiative. Retrieved from https://www.wcs.org/our-work/wcs-high-integrity-peatlands-cross-regional-initiative

World Bank. (2022). State and trends of carbon pricing 2022. https://documents.worldbank.org/en/publication/documents-reports/documentdetail/099045006072224607

World Bank. (2024). *State and Trends of Carbon Pricing: International Carbon Markets 2024*. Washington, DC: World Bank. http://hdl.handle.net/10986/42094

World Economic Forum. (2025, January). Bridging the gap: How to finance the net-zero transition. White Paper. https://www.weforum.org/publications/bridging-the-gap-how-to-finance-the-net-zero-transition/

World Wildlife Fund. (n.d.). Conserving wildlife and enabling communities in Namibia. https://www.worldwildlife.org/projects/conserving-wildlife-and-enabling-communities-in-namibia

Zhang, T., & Zhao, X. (2014). Do kinship networks strengthen private property? Evidence from Rural China. *Journal of Empirical Legal Studies*, 11(3), 505–540.

Zhu, A. L., Ndiaye, A., Dahm, R., Mauclaire, M., & Boas, I. (2025). Africa's Great Green Mirage? Assessing the disconnect between global finance and local implementation in Africa's Great Green Wall. Land Use Policy, 157, Article 107515. https://doi.org/10.1016/j.landusepol.2025.107515

6 Integration and Stakeholder Engagement in Climate Mitigation

A Logical Framework for a Paradigm Shift

Since 2005, REDD+ has prioritized top-down emissions reductions, often sidelining biodiversity, legitimacy, and IPLC governance. Despite initial promises of transformative change, REDD+ has too often reproduced established conservation paradigms, delivering continuity more than genuine reform (Lund et al., 2017). This approach has underperformed in remote, weakly governed landscapes, where IPLC priorities remain marginalized (Angelsen et al., 2018; Duchelle et al., 2023). It has also yielded disappointing results, especially in remote, biodiverse, and weakly governed landscapes where mitigation is most needed, while consistently sidelining IPLC priorities, despite IPLC customary tenure over more than one-third of the world's intact forests and receipt of less than 1% of global climate finance (Rights and Resources Initiative, 2018). Moreover, increasing evidence shows that a wide range of human activities – including degradation, fragmentation, and selective logging – are major drivers of tropical forest carbon loss, beyond outright deforestation (Pyles et al., 2022). Forests are also problematically defined, leading to perverse incentives and flawed restoration metrics (Chazdon et al., 2017).

To chart a more effective and just path forward, this chapter calls for a shift from carbon-centric, top-down approaches to context-specific integration that centers IPLC rights, knowledge, and governance and rethinks the very STAM used to pursue climate outcomes. Again, STAM refers to the full suite of frameworks and instruments, from policy tools to participatory methods, that shape how climate mitigation is designed and implemented in the land sector.

Over the past two decades, INRM and so-called inclusive climate action programs have laid useful groundwork – participatory negotiation, adaptive governance, and landscape-scale planning are now part of the mainstream lexicon (van Noordwijk et al., 2001). But despite the rhetoric of empowerment, mainstream climate finance still fails to prioritize the essentials: FPIC, robust IPLC capacity, and legal recognition of customary tenure remain the exception, not the rule (Brown & MacLellan, 2020; Conservation International, 2022; Duchelle et al., 2023; Luttrell and Pellini, 2025). INRM approaches that seek to explicitly address the water–energy–food–land nexus in context can help overcome the pitfalls of single-sector strategies (Herrera-Franco et al., 2023). Dooley et al. (2022) note the grave risks to food security, biodiversity, and Indigenous rights if tenure and ecosystem integrity are not prioritized in planning.

DOI: 10.4324/9781003380320-7

Programs such as UNDP's AIM and SCALA frequently emphasize "empowerment" yet sidestep the structural barriers – land grabbing, corruption, weak enforcement, and the persistent exclusion of IPLC women in climate action – that ultimately determine outcomes (Duchelle et al., 2023; SCALA/UNDP, 2025). As a result, "integration" is too often reduced to a technical exercise in cross-sectoral coordination rather than the kind of governance transformation that can actually shift power and legitimacy to the local level. Until these foundational gaps are addressed, climate action will remain less effective and less sustainable than it needs to be.

The following sections examine these challenges in practice, analyze operational lessons, and propose a new logic for integration grounded in IPLC rights, gender equity, and negotiated governance. This approach lays the foundation for the paradigm-shifting solution developed in Chapter 7, designed to deliver appropriate, feasible, and sustainable climate mitigation outcomes at scale.

Incentives, Mechanisms, and Capacities for Integration

As "locally led" and "integrated" approaches gain currency across climate and development discourse (see Box 6.1), the critical question is how these principles can be operationalized, especially in weakly governed, tenure-insecure contexts where mitigation is expected to expand.

Box 6.1 Proof of Concept: Selected Examples of Successful Integration and Its Limits

- **Ethiopia:** Persistent challenges, including a lack of stable cross-sectoral governance, economic priorities overriding climate goals, and weak information sharing, *hinder effective landscape restoration and resilience* (Bimir & Matus, 2025; Kassa et al., 2020).
- **Colombia:** Land-use conflicts fueled by overlapping claims and fragmented governance; participatory land-use planning post-peace accord reduced deforestation and strengthened community governance (O'Brien, 2025; Del Rio Duque et al., 2022).
- **DRC:** REDD+ projects focused on carbon, sidelining adaptation and biodiversity (NDC Partnership, 2024).
- **Brazil:** Amazon Soy Moratorium reduced deforestation on paper but failed to address tenure or Indigenous governance (Human Rights Watch, 2019).
- **Kenya:** Kasigau Corridor carbon finance undermined by revenue siphoning and mistrust (SOMO & Kenya Human Rights Commission (KHRC), 2023); participatory GIS mapping resolved 44 land disputes (IFAD, 2023).
- **Ghana:** CREMA model struggled to scale due to a lack of national alignment (MDPI, 2020).
- **Nepal:** Hariyo Ban program restored tens of thousands of hectares and improved water security, but such successes remain rare (WWF, 2016).

A few emerging initiatives offer partial reference points, though none yet provide a proven model for fully integrated, IPLC-centered climate mitigation. TNC's The Voice, Choice and Action Framework (2022) offers its vision for how BINGOs can work more effectively with IPLCs, without fully shifting authority to them.

The World Bank's Locally Led Climate Action (LLCA) initiative aspires to reach 100 million people by 2030 through participatory planning, institutional resilience, and direct community financing. It promotes climate diagnostics rooted in local knowledge, adaptive management, and fiscal transparency – principles that echo integrated governance. However, as discussed above, technocratic and fragmented approaches remain a barrier: Both LLCA and SCALA have yet to demonstrate how their designs ensure IPLC co-design, tenure protections, or negotiated governance. Lah (2025) offers a logical integration framework for addressing the kinds of disconnects LLCA has faced through adaptive co-design.

To move beyond conceptual rhetoric, integration must be operationalized through grounded models that align incentives, build IPLC capacity, and adapt to local contexts, no less so with how the classic challenge of gender integration is approached (Locatelli et al., 2015).

Gender: Shifting From Tokenism to Power-Sharing

Gender equity in climate mitigation cannot be achieved through token participation or rhetorical commitments. Lasting impact requires enforceable land and resource rights for women, guaranteed leadership roles in decision-making bodies, and finance tied to measurable outcomes – such as the percentage of project funds controlled by women's groups or legal recognition of women's tenure. Traditional ecological knowledge held by women must also be institutionalized in project design, with government and donor staff required to learn from women's leadership. With backlash against DEI programming in the United States and elsewhere (Box 6.2), leadership on gender issues is needed more than ever.

Box 6.2 The Global Backlash Against DEI and Its Impact on Climate and Gender Initiatives

Recent years have seen a sharp political backlash against Diversity, Equity, and Inclusion (DEI) policies across major donor countries, with direct consequences for international climate, gender, and development programs. As discussed in Chapter 5, DEI has become both a strategic imperative and a lightning rod in public discourse. In the United States, the 2025 administration's budget proposal slashed foreign assistance by 84%, explicitly targeting programs labeled as "radical, leftist priorities," including those focused on gender and equity (Devex, 2025; Jansen, 2025).

The United Kingdom has announced a 40% cut to its aid budget by 2027, redirecting funds from development to defense, while Australia and several European donors have also reduced official development assistance (Devex, 2025; Lowy Institute, 2025).

This retrenchment is partly fueled by populist rhetoric that frames DEI as "woke" or ideologically driven, despite mounting evidence that inclusive approaches are essential for effective, sustainable outcomes in both climate and development (FT Longitude, 2025). The paradox, as explored in Chapter 5, is that while resistance to DEI is rising, the operational need for gender equity and social inclusion has never been greater.

As aid budgets shrink and political polarization intensifies, the future of gender, equity, and climate initiatives in the Global South is increasingly uncertain, making it all the more vital that new governance models, such as NGPs, embed DEI as a foundational, negotiated principle rather than an optional add-on. Communicating *why* will be key.

Minimum Requirements for Genuine Gender Integration in NGPs

- Enforceable land and resource rights for women
- Guaranteed leadership roles and decision-making power
- Finance tied to measurable, gender-specific outcomes
- Institutionalization of women's traditional knowledge in governance

In the NGP model, gender power-sharing is a structural requirement, not an add-on: Agreements must codify women's rights and leadership, with credible metrics to track gender-specific outcomes, including veto authority over projects affecting their territories or livelihoods.

While some critics frame support for women's leadership as "ideological," the evidence is clear: Women are the primary stewards of land, water, and biodiversity across the Global South, and their substantive inclusion is essential for appropriate, feasible, and sustainable climate outcomes (CIF, 2024; Deininger et al., 2023). Embedding these provisions in NGPs is not a matter of ideology but a pragmatic response to operational realities on the ground.

Negotiated Governance Platforms: A Collaborative Solution

Traditional models of project and donor relationships have typically relied on rigid, top-down accountability structures. In these frameworks, demands for accountability are often unilaterally defined by donors or external actors, with limited input from local stakeholders. While such arrangements may ensure compliance with donor requirements, they frequently fail to foster genuine partnership, shared ownership, or context-sensitive solutions. These classic approaches risk perpetuating power imbalances and can undermine the very objectives they seek to achieve.

The limitations of the project/donor/classic model highlight the urgent need for a paradigm shift in how accountability and relationships are structured in international development and cooperation. Rather than imposing externally determined standards and requirements, there is a growing recognition that accountability must be co-constructed, negotiated, and adapted to the specific contexts and needs of all stakeholders involved.

To address the persistent failures of top-down and tokenistic "participation" in climate mitigation, this chapter proposes the establishment of NGPs. The concept builds on the types of success and proof of concept illustrated in Box 6.1.

NGPs are structured forums where the principal actors, including IPLCs, governments, donors, and the private sector stakeholders, jointly negotiate the terms of climate action, including tenure, benefit-sharing, and accountability frameworks. This approach draws on the concept of negotiated governance, which has gained traction as a means to overcome the inadequacies of hierarchical, siloed, or purely consultative models in sustainability (Sustainability Directory, 2025). Unlike conventional multi-stakeholder platforms that often remain advisory, NGPs are intended as operational, decision-making bodies embedded within national and subnational governance systems.

Unlike conventional multi-stakeholder platforms, NGPs are proposed as operational, decision-making bodies with the potential for binding authority, legal codification, and direct allocation of finance and power to IPLCs. While existing models have often been limited to advisory roles or consultative processes, NGPs would, if implemented, grant IPLCs veto rights over key decisions, require legal recognition of negotiated outcomes, and channel funding directly to community-controlled institutions. This distinction is critical: Previous models failed to deliver systemic change because they lacked enforceable mandates, left accountability with donors or governments, and treated IPLC participation as consultative rather than co-governing.

NGPs aim to institutionalize genuine power-sharing and accountability to IPLC constituencies. Table 6.1 summarizes how NGPs differ from conventional approaches and what is needed to operationalize them across diverse contexts. NGPs aim to institutionalize genuine power-sharing and accountability to IPLC constituencies.

One of the key criticisms of NGPs is the risk of elite capture – where dominant stakeholders, including powerful state actors or corporate interests, may co-opt NGP platforms for their own benefit. To mitigate this risk, independent oversight mechanisms, such as third-party monitoring and community-led evaluation processes, must be integral to the NGP structure. These mechanisms should be embedded from the outset, ensuring that accountability flows directly to IPLC communities and not just to donors or governments.

Table 6.1 NGPs Compared to Other Frameworks

Framework	*Tenure Rights*	*IPLC Co-Governance*	*Gender Inclusion*	*Adaptation Focus*
REDD+	Weak	Limited	Low	Low
Carbon Pricing	Weak	Limited	Low	Low
INRM	Stronger	Moderate	Moderate	Moderate
NGPs	Strong	Strong	Strong	Strong

To mitigate the risk of elite capture, NGPs must include clear safeguards such as independent third-party monitoring and community-led evaluation mechanisms. For example, an independent evaluation body, consisting of external auditors and representatives from IPLC organizations, can regularly assess the implementation of agreements to ensure that the power-sharing model is respected and upheld. Additionally, community-led monitoring systems – such as those used in Liberia's Land Rights Act – can play a key role in identifying abuses of power and ensuring that resource distribution is equitable. It is essential to embed these safeguards in the governance architecture from the outset to ensure accountability flows directly to IPLC communities and not to intermediaries or powerful state actors, who have historically marginalized these communities.

While NGPs are offered as a solution to top-down climate governance, their true test will be whether they deliver enforceable shifts in power and resources to IPLCs through negotiation with the government and the private sector. The greatest risk is that NGPs become vehicles for co-option or "business-as-usual" under a new label.

To avoid this, NGPs must be held to a higher standard: IPLCs must have measurable control over decisions and budgets, legal authority must be codified, and accountability must flow to IPLC constituencies, not just to donors or governments, as has been previously pointed out (Green Climate Fund, 2022). NGPs should be independently evaluated by empowered, resourced IPLC organizations (WRI, 2023), with clear mechanisms to prevent dilution, reversal, or greenwashing. Only by embedding these safeguards can NGPs serve as a genuine basis for a paradigm shift in climate governance.

Table 6.2 presents further detail for the rationale and precedents for NGPs, the conditions for giving them a fair pilot in diverse national and subnational contexts, and the accompanying actions needed to enable forthright negotiations leading to appropriate and feasible outcomes for people, the environment, and ultimately the climate to be reached.

While NGPs emphasize negotiation and contextual adaptation, they do not disregard established best practices or international standards. Instead, these serve as a starting point or reference, guiding the negotiation process and helping to ensure that fundamental principles, including transparency, the legal basis for FPIC (Doyle & Cariño, 2013), and equity, are upheld.

The shift from classic project/donor accountability models to NGPs represents a fundamental rethinking of how relationships and responsibilities are structured in development cooperation. By prioritizing negotiation, inclusivity, and contextual adaptation, NGPs offer a pathway to more effective, equitable, and sustainable outcomes.

Operationalizing Integration: From Principle to Practice

Despite growing consensus on the rationale for integrated, IPLC-centered climate mitigation (Rainforest Foundation US, 2024), most climate and development policies remain more rhetorical than operational. To move beyond pilots and slogans, climate actors must focus on building institutional systems that embed integration in law, finance, and governance structures – making it real, durable, and scalable (Bimir & Matus, 2025; van Noordwijk et al., 2001).

Table 6.2 From Siloed Action to Integrated Governance: The NGP Paradigm Shift

Negotiated Governance Platforms (NGPs) are structured forums where Indigenous Peoples and Local Communities (IPLCs), governments, donors, and other stakeholders jointly negotiate the terms of climate mitigation, including tenure, benefit-sharing, capacity building, and accountability, rather than having these imposed from above. NGPs institutionalize shared decision-making, moving beyond token IPLC engagement toward genuine integration and negotiated co-governance (Bäckstrand, 2006).

Why a Paradigm Shift?

Current climate mitigation remains fragmented:

- **Institutional silos**: Ministries and agencies operate with conflicting mandates.
- **Donor fragmentation**: Less than 1% of climate finance reaches IPLCs directly, creating patchwork projects (Altamirano et al., 2022).
- **Market myopia**: Carbon markets often sideline local legitimacy and stakeholder engagement.

Evidence for Negotiated Integration

Collaborative planning pilots, such as Colombia's participatory land-use reforms, Namibia's conservancies, Kenya's community mapping, and Nepal's community forests (see Box 6.5), demonstrate that IPLC empowerment and shared planning yield better outcomes. However, these are not yet full NGPs, nor have NGPs been tested or proposed at scale. They serve as proof of concept for the potential of negotiated, integrated approaches.

Why Now?

- **Failed models**: Market-centric and top-down approaches have underdelivered for over a decade.
- **Cost-effectiveness**: Negotiated platforms can reduce leakage and misaligned incentives.
- **Global momentum**: The Paris Agreement and SDGs prioritize inclusive, integrated action.

NGP Blueprint

Enabling Condition	**Action to Be Taken**
Political Will	Embed NGPs in national climate strategies
Legal Recognition	Secure IPLC tenure; codify FPIC as binding
Community Capacity	Invest in IPLC analytical and negotiation capacity building
Adaptive Financing	Pool funds for milestone-based, flexible support

Incentives for Participation

- **Governments**: Reduced conflict, unlocked finance, SDG alignment.
- **Private Sector**: ESG credibility, de-risked investments, new markets.

Conclusion

NGPs represent an evidence-based, still partially untested, pathway to move from fragmented pilots to systemic, scalable integration. Centering IPLCs as co-architects offers the best chance to align climate action with sustainability. Negotiation Governance Platforms (NGPs) are proposed as a central innovation. While past collaborative planning pilots provide encouraging precedents, NGPs themselves have *never* been implemented or evaluated at scale with the intention of bringing IPLCs, government, and the private sector to the table. They should be broadly tested, as they offer the highest potential to break out from the current dependency on external "expert"-driven CMLSGS design and implementation. Unless negotiation becomes the core standard, appropriate, feasible, and sustainable climate mitigation will not be attainable.

Institutionalizing Integration Across Sectors and Scales

Operationalizing integration means embedding it in law, finance, and governance across ministries and levels of government. This requires:

- Cross-sectoral governance bodies with real decision-making power
- Joint planning and budgeting for climate, biodiversity, and development
- Legal frameworks that mainstream integration, not just silo policies (See Table 6.3 for a summary of operational steps.)

Box 6.3 Approaching Constraints to Implementing NGPs in Weak Governance Contexts

Key Challenges

Elite resistance: Powerful interests block reforms that threaten their control over land or resources.

Low state capacity: Weak institutions struggle to enforce agreements or deliver services.

Political economy fragmentation: Ministries and donors often pursue competing agendas, undermining integrated governance.

Options for Scaling and Adapting NGPs

Embed NGPs in national law and climate strategies to ensure continuity beyond political cycles.

Mandate FPIC and secure IPLC tenure rights.

Decentralized, Adaptive Pilots

Launch NGPs where IPLC organizations are strong and customary tenure is recognized.

Build from local legitimacy upward, linking customary and formal governance.

Require transparency in land and finance flows.

Channel climate finance directly to IPLC-managed funds, bypassing intermediaries.

Align donor funding with community priorities and public goods.

Use flexible, milestone-based financing and iterative learning.

Link private sector and elite benefits (e.g., ESG access, reputational gains) to verified NGP participation and compliance.

Safeguards Against Backsliding

Establish grievance mechanisms and independent mediation.
Create community-controlled trust funds to maintain momentum during political transitions.

Conditions for Success

Start with subnational pilots where IPLC capacity is high.
Use transparency, third-party monitoring, and adaptive financing to mitigate risks.
Align incentives across government, donors, and private actors.

Lessons Learned

Avoid over-reliance on carbon markets or one-size-fits-all models.
Prioritize legal empowerment, negotiated authority, and IPLC-led metrics for durable, scalable integration.

Table 6.3 Operational Steps

Step	*Why It Matters*
Cross-sectoral governance	Breaks silos and enables joint planning and decision-making at different levels
Negotiated Governance Platform (NGP) planning and budgeting	Aligns incentives and shared accountability
Potential for establishing legal codification	Takes integration from a project-based activity to a structurally sustainable mechanism
IPLC tenure security	Foundation for stewardship and legitimacy
Participatory metrics collaboratively established	Ensures accountability to IPLC priorities that enable sustainability
Capacity building	Enables negotiation, monitoring, and adaptive learning
Adaptive management	Enhances the probability that NGPs work

As summarized in Table 6.3, operationalizing integration requires embedding negotiated governance, IPLC leadership, and participatory metrics into law, finance, and institutional practice. This means securing tenure, codifying FPIC, building IPLC capacity, and ensuring accountability flows to IPLC constituencies, not just donors or governments.

Financing Integration at Scale: Weaning From Donor Dependency to Measured Support

Despite growing recognition of IPLCs' stewardship, less than 1% of global climate finance reaches IPLC-governed organizations, and most of this is still funneled through large intermediaries or project-based grants (Forest Declaration Assessment, 2024; Rainforest Foundation Norway, 2021; World Bank Development Finance Vice Presidency, 2022). For NGPs and integrated approaches to succeed, finance must be diversified, predictable, and negotiated – flowing directly to IPLC institutions and reflecting the priorities of all three parties: government, IPLCs, and the private sector, including philanthropists and impact investors (IMF, 2022).

This tripartite model will rely to some degree on:

- **Blended finance:** Combining public (national/international), private sector, and philanthropic capital, using innovative instruments such as blended funds, guarantees, and risk-sharing to de-risk investments and attract private and impact capital (Forest Declaration Assessment, 2024; IMF, 2022).
- **Direct access:** Creating direct-access windows for IPLC-led initiatives within all major funding streams – not just donor or government programs – and simplifying administrative barriers for local entities (Crawford & Church, 2019; Rainforest Foundation Norway, 2021).
- **Negotiated milestones:** Linking disbursement from all sources – public, private, and philanthropic – to transparent, negotiated milestones and metrics set by the NGP, not just donor reporting cycles (Forest Declaration Assessment, 2024).
- **Transparency and tracking:** Adopting transparent tracking and reporting, such as OECD-DAC markers and the Path to Scale Dashboard, to ensure funds reach IPLC priorities and are not lost in intermediaries or bureaucracy (RRI and Rainforest Foundation Norway, 2024).

Without this financial shift, anchored in NGPs and blended capital, integration will continue to risk remaining rhetorical. The NGP model is designed to purposefully align incentives, manage risks, and ensure that all parties, including government, IPLCs, and the private sector/philanthropy, have both a perceived and real stake and voice in both the negotiation and the flow of resources, as has repeatedly been noted as needed (Forest Declaration Assessment, 2024; IMF, 2022).

Metrics and Integration

The following examples in Box 6.4 show how integrated, negotiated metrics are being applied in diverse contexts.

Box 6.4 also demonstrates that integration can work when metrics align with IPLC priorities, are embedded in legal frameworks, and are supported through institutional investment. Yet they remain fragile – limited in scope, donor-dependent, and vulnerable to policy reversals. The real test is whether climate governance can

Box 6.4 Applying Metrics of Integration in Practice

Examples from Ghana, East Kalimantan, and Guatemala

Metric of Integration	*Real-World Application*
1. IPLC-Led Diagnostics and Planning	In Guatemala's Maya Biosphere Reserve, IPLCs conduct forest inventories, define management rules, and co-develop concession plans with state forestry authorities (Rainforest Alliance, 2023).
2. Multi-Scalar Consent Mechanisms	Indonesia's East Kalimantan J-REDD pilot included provincial, district, and village-level engagement, but later critiques noted weak FPIC verification (Forest Peoples Programme, 2024).
3. Tenure Recognition Benchmarks	Ghana's CREMAs offer formal legal status to community conservation efforts, anchored in national legislation and linked to district-level governance (UNDP, 2024).
4. Negotiated Benefit-Sharing and Transparency	CREMA communities codify revenue-sharing rules into bylaws covering tourism and carbon programs, with joint monitoring of benefits and obligations (UNDP, 2024).
5. Customary Governance Embedded in Law	Forest concessions in Guatemala legally enshrine customary decision-making councils, with contract renewal contingent on compliance with both forest health and governance metrics (Cronkleton et al., 2011).

shift from pilot improvisation to become a roadmap for broader systemic legitimacy, one that is rooted in negotiated authority and multi-stakeholder accountability.

The deeper challenge is whether these models *can* be scaled and sustained or whether their success is inherently constrained by the very paradigm they operate within. If integration continues to depend on exceptional cases, ad hoc partnerships, and external subsidies from Overseas Development Assistance (ODA) funding or carbon markets, then it cannot and will not serve as the foundation for systemic transformation or even, within the framework of the current paradigm, sustainability.

This is where the paradigm shift proposed in this book becomes essential. The path forward is not to discard current models but to elevate them into institutional norms. The goal is not to abandon donor-funded pilots, but to build national and subnational systems that formalize their principles: NGPs, codified consent mechanisms, legal tenure recognition, and community-defined indicators of success.

Only by moving beyond improvised integration toward structural, scalable legitimacy can climate mitigation efforts in the land sector achieve both equity and durability. The next section explores how this effort is complicated – and often undermined – by the dominant narratives of scientific urgency, efficiency, and market rationality that continue to shape climate finance and programming in the Global South.

Science, Urgency, and the Risk of Expedient Action

In the current era of climate politics, *the science* is often invoked not as a guide to feasible action but as a rhetorical cudgel (Pielke, 2007). Politicians and donors frequently proclaim that "the science is clear" to justify ambitious funding targets and compressed implementation timelines, often with little meaningful engagement of the communities expected to carry the burden of these strategies. While FPIC is often check-boxed as having been achieved, the merits of how genuine the FPIC is have been frequently questioned (Anaya, 2009; Brown, 2013; Proforest & Landesa, 2023; Sawathvong & Hyakumura, 2024). What began as a movement grounded in scientific assessment has, in many cases, hardened into a discourse of inevitability, one that leaves minimal room for political deliberation, feasibility constraints, or local governance priorities (Hulme, 2009; Turnhout et al., 2016).

This disjuncture is starkly visible in the Global North. For example, a 2024 Pew survey found only 31% of Americans prioritize federal climate action, with deep partisan divides (Kennedy & Tyson, 2024), while European support for climate measures declines when associated costs rise (European Commission, 2023; Fairbrother, 2022).

Meanwhile, political backlash intensifies. The Trump administration's 2025 bid to gut the National Oceanic and Atmospheric Administration NOAA climate programs (Voosen, 2025) exemplifies how science has become a partisan battleground, undermining global cooperation. To justify expediency, recent critiques of the social cost of carbon (SCC), a key tool in determining the economic rationale for mitigation, reveal troubling trends. While Nobel Laureate William Nordhaus proposed relatively conservative estimates (Nordhaus, 2017), Bilal and Känzig's (2024) revised SCC estimates are 10–20 times higher than Nordhaus' who recently noted that current policies will likely see the global temperature surpass the 1.5 °C limit this decade and exceed 2 °C by mid-century unless major changes are made, jeopardizing the targets of the Paris Agreement (Nordhaus, 2024). These risks I would argue, influence justify top-down interventions in the Global South. Unless paired with enhanced transparency and stakeholder deliberation offered by NGPs, this programming will be unlikely to succeed.

At the same time, calls for urgent climate action continue to escalate in international forums. Some donor narratives increasingly suggest that "doing something" in the Global South – such as launching large-scale tree planting schemes, jurisdictional carbon deals, or biodiversity offsets – is preferable to confronting entrenched fossil fuel interests or lifestyle transitions at home (Ecosystem Marketplace, 2025; IEA, 2025; Lang, 2025; Stechemesser et al., 2024). This approach treats the Global South as a laboratory for rapid action while counterproductively sidestepping feasibility, consent, or justice (Harvey & Schonhardt, 2024; Lomborg, 2021). While many see UNFCCC's Article 6.4 as fundamental, others see it as expanding unproductive and harmful BAU (see Lang, 2024).

Instead, what is needed is participatory science that clarifies trade-offs and supports negotiated solutions, not dogma (Bustamante et al., 2014a; Turnhout et al., 2013; van Bommel & Höffken, 2023). One possible model to reconsider for trade-off analysis that was highly successful at participatory land-use planning, tested with proof of concept over a seven-year period under USAID funding in Cameroon and the DRC was the Community Options Analysis and Investment Toolkit (COAIT), which empowers communities in remote landscapes to generate ecological, economic, and social data as part of participatory cost-benefit and risk analysis for sustainable natural resource management (Bonis-Charancle et al., 2009).

Hulme (2009) argues that climate science must be re-situated as a contributor to pluralistic, place-based conversations, not as a hammer to close down political space. For IPLCs, this requires tools and data to independently assess project risks/benefits, ensuring science informs, not overrides, local decision-making. Here, tools like a modernized COAIT are logical to revisit.

This erosion of trust in scientific institutions mirrors declining public confidence in media and governance more broadly (Gallup, 2024). It highlights the failure of top-down urgency narratives to build public legitimacy – and raises questions about whether the science-driven logic of urgency is politically sustainable in donor countries.

This urgency–justice dilemma (van Bommel & Höffken, 2023) manifests in the Global South as technocratic metrics and deadlines misaligned with local governance. Ultimately, reconciling scientific urgency with justice requires transparent, inclusive processes that center IPLC agency, a theme explored further in the final section on reconciling markets, feasibility, and equity.

Reconciling Justice, Markets, and Feasibility

CMLSGS requires an integrated approach that reconciles three core domains: justice, market efficiency, and feasibility. While each has received attention in isolation, efforts to integrate them remain fragmentary and conceptually underdeveloped. Too often, they function as parallel universes of concern – justice advanced by IPLC rights advocates, markets championed by donor finance and private investors, and feasibility treated as a technical domain by government planners or multilateral agencies. Justice is framed through rights-based and "just transition" agendas (Climate Horizons, 2024; Laudes Foundation et al., 2024); markets are mobilized through instruments like REDD+ and Article 6 carbon trading (Ecosystem Marketplace, 2023); and feasibility is treated as a technical afterthought, rarely subjected to political or cultural analysis.

The challenge is not merely to balance these priorities but to redefine the terms of integration in light of political polarization, governance capacity, and the lived realities of IPLCs. In today's hyper-politicized landscape, especially given budget austerity, climate denialism, and populist backlash in many donor countries, only the most appropriate and feasible strategies should be advanced. Ambitious targets unsupported by socially legitimate methods risk undermining both justice and impact.

The historical record offers sobering lessons. Despite over a decade of REDD+ programming and NbS, large-scale emissions reductions and biodiversity conservation remain elusive. In many cases, these approaches have been more effective at generating hype than beneficial outcomes to both people and planet (Duchelle, et al., 2018; Massarella et al., 2018; Yazzie, 2022). The disconnect stems from fundamental misalignments between the normative frameworks of the international climate community and the political and governance systems in which these programs are implemented, particularly in landscapes managed under customary tenure (Brown & MacLellan, 2020; Rights and Resources Initiative, 2022).

Market-based mechanisms like REDD+ and voluntary carbon offsets have struggled with elite capture, weak grievance systems, and limited local participation – raising persistent concerns about integrity and inclusion (Worsdell & Sambhav, 2020). IPLCs are rarely positioned as co-equal actors with the capacity to negotiate or adapt programs to their governance systems (Duchelle et al., 2023). Where IPLC-led models have been supported, the results are more durable, equitable, and ecologically effective (Garnett et al., 2018; Rights and Resources Initiative, 2023).

Achieving a genuine paradigm shift in CMLSGS will not come from optimizing existing frameworks. There is no single technocratic model that can square these competing logics. Rather, multiple logics and illogical positions must be negotiated in place-based, iterative ways. Attempts to apply universal incentives or neoliberal models to drive large-scale behavior change have repeatedly failed.

Yet international climate finance debates remain stuck in a misdiagnosis: assuming that the central problem is a lack of money. While finance is indeed a bottleneck, pouring billions into REDD+, Article 6-linked carbon markets, or fragmented AFOLU initiatives without confronting structural feasibility problems is unlikely to yield impact – a logic often reinforced by economic discounting that devalues long-term social and ecological outcomes (Barrage & Nordhaus, 2024). Similarly to Brown's (2013) argument over a decade ago, failure to invest in obtaining social license, negotiated governance, and community empowerment means that even well-intentioned funding will backfire or be wasted.

IPLCs in CMLSGS

IPLCs are indispensable to effective and sustainable climate mitigation in the Global South. Their stewardship of critical ecosystems – accounting for nearly 40% of the world's remaining intact landscapes – underscores their vital role in maintaining biodiversity, ecosystem services, and carbon sinks (Garnett et al., 2018; Inclusive Conservation Initiative, 2022). Numerous studies confirm that IPLC-managed lands exhibit lower deforestation rates, stronger biodiversity outcomes, and greater long-term resilience than many government-protected or privately managed areas (Rights and Resources Initiative, 2023; World Resources Institute, 2024).

Yet IPLCs continue to be marginalized in global mitigation frameworks. Too often treated as passive beneficiaries rather than active agents, their rights, governance systems, and land claims have been sidelined in favor of externally designed, market-oriented approaches (Cultural Survival, 2022; Hunsberger et al., 2017; Rights and Resources Initiative, 2023). The result has been widespread underperformance of mitigation programs like REDD+, which frequently fail to engage communities meaningfully or deliver sustainable, equitable results (Brown & MacLellan, 2020; Carbon Market Watch, 2024; Massarella et al., 2018).

The political economy of land continues to undermine IPLC empowerment. In many countries, land allocation systems still favor elites, extractive industries, and speculative carbon finance arrangements – often at the expense of community rights and long-term stewardship (Rights and Resources Initiative, 2020b; Sunderlin et al., 2018). This reflects a deeper contradiction in the prevailing mitigation logic: the assumption that behavior change can be induced through commodified carbon incentives without addressing the foundational question of who holds decision-making authority over land.

As this book has argued, the failure to reconcile governance, legitimacy, and local feasibility has undercut most large-scale mitigation efforts to date. A new paradigm must move beyond reductionist, technocratic, and transactional models, where more protected areas or prioritization of infeasible carbon-reduction projects are favored, as the social costs (Ferraro & Hanauer, 2014) are simply too high. It must prioritize IPLC knowledge systems, negotiated governance arrangements, and diverse incentive structures that reflect the realities of place (Pascual et al., 2021; Tauli-Corpuz et al., 2020). Market-based mechanisms alone are insufficient. What is needed is a redefinition of success, grounded in co-benefits, inclusive governance, and IPLC-determined priorities.

Lessons From Practice to Build Upon

To ground the preceding conceptual arguments in real-world experience, this section examines how integration and stakeholder engagement have played out in practice. Drawing on examples from Colombia, Nepal, and Namibia, it highlights distinct pathways for breaking silos, sequencing reform, and rebalancing power in climate mitigation. Each example offers practical lessons for designing more integrated, legitimate, effective, and durable approaches in the land sector of the Global South.

Across these diverse contexts, three catalysts have repeatedly proven effective in driving paradigm shifts, even in settings marked by weak governance: crisis as opportunity, the power of cross-sector coalitions, and fiscal pragmatism that earns legitimacy. The following case studies illustrate how these catalysts have enabled durable, IPLC-centered climate mitigation.

Breaking Silos and Legal Reform: Colombia's Peace Accord

Acute shocks – whether post-conflict transitions, climate disasters, or economic collapse – can create political openings for systemic reform. Colombia's 2016 peace accord exemplifies this dynamic. By leveraging the urgency of post-conflict reconstruction, the government embedded participatory land-use planning into national law, recognizing Indigenous and Afro-Colombian territories. This legal recognition enabled participatory land-use planning, reducing deforestation by 34%. It achieved this while strengthening community governance and institutionalizing IPLC-led conservation as a pillar of rural development (NDC Partnership, 2024). Crisis-driven coalitions, such as those formed during Colombia's post-conflict reforms, created political space for collaboration among IPLCs, the state, and donors. The Colombian case demonstrates how reframing crisis as an opportunity for governance renewal, rather than a mere technical fix, can break down institutional silos and establish new norms for integrated, rights-based climate action – even as implementation remains uneven and faces persistent gaps (O'Brien, 2025).

Sequencing Reform and Community Leadership: Nepal's Community Forestry

Nepal's community forestry movement demonstrates how multi-decadal legal reform, adaptive financing, and IPLC capacity building can institutionalize local leadership at scale (see Box 6.5) (Agrawal & Ostrom, 2001).

Box 6.5 Nepal's Community Forestry as a Paradigm-Shifting Model

Nepal's community forestry is arguably the most successful and scalable example of IPLC-led, integrated governance – yet it remains underutilized in global climate discourse. Key revisions to highlight its relevance:

Nepal's Community Forestry – A Blueprint for NGPs:

- Scale: 22,266 Community Forest User Groups (CFUGs) manage 34% of Nepal's forests, directly benefiting 2.9 million households (See FECOFUN homepage at https://fecofun.org.np).
- Ecological Impact: Forest cover increased from 39.6% to 44.7% (1990–2015) in CFUG-managed areas, outperforming state-protected zones (Nepali Times, 2019).
- Governance Innovation:
  - Iterative legal reforms (1976–1993 Forest Acts) embedded customary systems into national law.
  - Revenue retention: CFUGs retain 100% of forest product revenues, reinvesting in schools, healthcare, and microenterprises (World Bank Group, 2022).
  - Conflict resolution: Federations like FECOFUN mediate disputes and advocate for policy change, reducing elite capture (Ghimire, 2024).

Why Nepal Matters for NGPs:

- Demonstrates how multi-decadal legal/policy sequencing can institutionalize IPLC leadership.
- Shows that pooled, adaptive financing (e.g., Hariyo Ban's $50M fund) aligns with community priorities better than project-based aid.
- Illustrates that IPLC capacity building (e.g., GIS mapping, legal literacy) enables self-governance to be achieved at scale, though challenges remain (Dhungana et al., 2024).

Fiscal Incentives and Power Rebalancing: Namibia's Conservancies

Namibia's conservancies align incentives with IPLC priorities, demonstrating that hybrid governance and revenue sharing can drive biodiversity conservation success (see Box 6.6) in favorable situations even with resistant elites who may back reforms when faced with clear cost-benefit incentives, and where land use conflict is limited. This,, in addition, to land tenure security (Reytar et al., 2024).

Adapting Global Climate Governance for IPLC Leadership

The legitimacy and effectiveness of CMLSGS hinge on transforming global governance systems, especially those that shape policy norms and finance flows. Institutions like the UNFCCC and IPCC still reflect a legacy of top-down, state-centric models that exclude IPLCs from meaningful participation in rulemaking, program design, and finance allocation. At best, IPLCs are consulted; at worst, they are bypassed entirely.

To meaningfully shift this paradigm, international climate governance must structurally embed IPLC leadership. Several reforms could enable this:

1. Establish Permanent IPLC Representation: Institutions such as the UNFCCC should allocate formal, decision-making seats to IPLC representatives. While

Box 6.6 Namibia's Conservancies: Aligning Incentives for Conservation

- 90% of tourism and hunting revenues go to communities.
- $166 million generated annually
- Over 1,000 full-time jobs created
- N$107 million contributed to the national economy in 2023

Supported by WWF and GIZ, reduced poaching and habitat loss (NACSO, 2024; WWF Namibia, 2023).

the Local Communities and Indigenous Peoples Platform represents a step forward, it lacks enforcement power. A reformed governance model could include regionally diverse IPLC seats with veto rights over land-sector projects that affect their territories.

2. Enforce Robust FPIC Standards: FPIC must be redefined not as a procedural formality but as a substantive, enforceable right. Project financing should be contingent upon independent verification of consent, derived from processes that are transparent, representative, and informed. This requires building IPLC capacity to understand, negotiate, and monitor agreements – and holding governments and intermediaries accountable for violations (Conservation International, 2022; Global Dialogue, 2023).
3. Broaden Metrics Beyond Carbon: International frameworks must recognize that IPLCs often deliver climate mitigation through stewardship practices that also support biodiversity, cultural integrity, and food sovereignty. Developing and institutionalizing metrics that account for these non-carbon benefits can help realign global targets with local priorities (Ford Foundation, 2025; WRI, 2023).

Without these reforms, climate governance will continue to reproduce the same structural exclusions that have plagued mitigation efforts for over a decade. Moving forward, IPLC leadership must be treated not as an aspiration but as a precondition for climate success.

Hybrid Governance Models and Co-Designed Platforms

IPLCs into climate governance requires models that move beyond conventional state or market paradigms. Hybrid governance – anchored in both customary systems and statutory frameworks – offers a promising path forward, but only when backed by enforceable rights, institutional accountability, and sustained participation. These models are not about compromise for its own sake; they are about recalibrating climate governance to reflect plural legal systems, shared authority, and negotiated legitimacy.

Indonesia's *Kesatuan Pengelolaan Hutan* (Forest Management Units, or FMUs) provide one such institutional experiment. FMUs were designed to create participatory spaces where IPLCs could directly engage state agencies and private actors.

South–South peer learning has also emerged as a powerful tool for co-designing governance frameworks. The Pan-African Network of Community Land Trusts builds IPLC capacity by fostering legal literacy, negotiation strategies, and regional solidarity (LandMark, 2024). These platforms offer not only practical skills but also political imagination, helping IPLCs reframe themselves not as project beneficiaries but as sovereign actors in climate governance.

On the private-sector side, Akaysha Energy's (2024) Social License Framework in Australia reflects a growing effort to embed Indigenous knowledge and oversight into renewable energy development. Bolivia's *Ley Marco de la Madre Tierra* anchors Indigenous worldviews and Buen Vivir principles in national climate finance, promoting community stewardship and support for agroecological initiatives. While the law does not mandate a specific share of carbon revenues for communities, its governance framework prioritizes inclusive, rights-based climate action (EcoJurisprudence Initiative, 2023).

To support such models, the Rocky Mountain Institute has proposed financial mechanisms designed to protect IPLC rights and limit extractivism. These include treating carbon sequestration as a time-bound service rather than a tradable asset, creating protective contract structures, and backing carbon-linked securities with real governance outcomes (RMI, 2023). In theory, such innovations could increase IPLC bargaining power. But in practice, without enforceable tenure, binding dispute resolution mechanisms, and IPLC co-design at every stage, even well-intentioned financial engineering risks reproducing the same asymmetries that have long defined market-led climate mitigation.

Ultimately, hybrid governance models are not simply technical fixes. They are expressions of a deeper shift toward negotiated legitimacy: a recognition that climate governance must be rooted in consent, accountability, and culturally grounded authority. These frameworks, or models, succeed when they redistribute power, which leads to greater balance, anchoring climate action in the institutions, relationships, and legal frameworks that IPLCs help define. In this sense, hybrid governance is not an end state but a process, a living architecture for the kind of paradigm shift this book envisions.

Table 6.4 provides a comparison of information about attributes for each framework or model. The Table highlights that most mainstream frameworks, such as REDD+ and carbon pricing, focus narrowly on carbon (Duchelle et al., 2018). They offer weak or limited tenure rights, limited co-governance, and little attention to gender or adaptation. INRM and NGPs, by contrast, provide stronger pathways for IPLC authority, gender inclusion, and adaptation. Their effectiveness depends on context and implementation. The persistent gaps, especially concerning enforceable rights, genuine power-sharing, and accountability to IPLCs, underscore the need for a paradigm shift toward integrated, locally anchored governance.

Moving from successful pilots to national systems requires legal codification of NGP mandates, predictable finance for IPLC institutions, and broad-based coalitions that include reformist bureaucrats, civil society, and private sector allies. Institutionalizing NGPs in law and national climate strategies, linking finance to verified IPLC outcomes, and building cross-sector alliances are essential to ensure that integration survives political turnover and donor cycles.

Scaling Integration in Weak Governance Contexts

Integration – the alignment of climate action across sectors, scales, and stakeholders – is often framed as a technical challenge. Yet even in high-capacity democracies like Norway, breaking institutional silos requires continuous negotiation (Oseland, 2019). In the Global South, weak rule of law, corruption, and elite competition magnify these barriers. If integration struggles in such contexts, the challenges are exponentially greater, as weak governance structures, corruption, and competing elite interests dominate. Technocratic "solutions" do not solve problems that are fundamentally political, economic, and cultural in nature.

Table 6.4 Governance and Equity Attributes Across Climate Mitigation and Land-Use Frameworks

Framework	*Goal Setting*	*Tenure Rights*	*Participation/ Co-Governance*	*Accountability*	*Gender Inclusion*	*Adaptation*	*Sectoral Scope*
REDD+	Carbon	Weak	Consultation, not binding	Donor/Market	Weak	Sometimes	Land/Forests
INRM	Eco/Social	Medium	Negotiated, context-specific	Variable	Variable	Yes	Land/Water
NGPs	Context-based	Strong	IPLC co-governance	Legal/Customary	Strong	Yes	Flexible
Carbon Pricing	Carbon	Not required	None	Market	None	No	Energy

Moving from pilots to scaled, integrated systems demands three key steps:

1. Legal Codification: Embed NGPs in national climate strategies and land laws. Liberia's 2018 Land Rights Act, which formalized customary tenure through donor-backed reforms, offers a template (World Bank, 2024).
2. Predictable Finance: Direct climate funds to IPLC-managed institutions, bypassing intermediaries. Namibia's conservancies, where 90% of revenues go directly to communities, demonstrate how revenue retention sustains conservation (WWF Namibia, 2023).
3. Coalitions for Continuity: Build alliances with reformist bureaucrats, civil society, and ESG-aligned businesses. Nepal's community forestry movement scaled nationally through grassroots legal advocacy and donor alignment (Dahal & Chapagain, 2005).

While these three key steps lay the groundwork for, ideally, expanding NGPs into formal governance systems, it is essential to also consider how to practically shift from pilot projects to fully integrated, operational systems.

However, as NGPs scale and transition to fully integrated systems, they must confront systemic barriers, particularly those entrenched in governance and financial mechanisms. These challenges are especially pronounced in countries where integration efforts collide with systemic fragmentation and corruption (ClientEarth, 2023; Freedom House, 2024), and where adaptation and mitigation efforts remain poorly coordinated, limiting the system-wide impact (IISD, 2021).

To facilitate the transition of NGPs from pilot projects to larger-scale systems, efforts ideally would focus on embedding them within national climate strategies, formalizing them in legislation, and aligning them with existing local governance structures where feasible. A manageable approach involves advocating for the gradual integration of NGP principles into national policy frameworks, starting with small-scale pilots and localized initiatives that demonstrate their value. Over time, these pilots can help build the political momentum and institutional capacity needed for broader adoption.

To support the successful scaling of NGPs, financial models must prioritize direct, unrestricted funding for IPLC-led initiatives. This can be achieved by creating dedicated funding windows within major climate finance streams, such as the Green Climate Fund or the Global Environment Facility, specifically for IPLC-managed projects.

Additionally, blended finance mechanisms, which combine public, private, and philanthropic capital, should be utilized to reduce risks and attract greater investment from impact investors and corporate donors. These blended funds should be tailored to local governance needs, with financial decision-making power delegated to IPLC leaders, thus ensuring that funds are spent in line with community priorities. Moreover, financial models should ensure that adaptive financing is embedded, where disbursements are linked to milestone-based, community-verified outcomes, not just donor-imposed metrics.

In addition to solid financial structures, collaboration with key stakeholders, such as national governments, local authorities, donors, and private sector actors, is critical for aligning incentives and securing the support necessary to expand NGPs. Through this phased, adaptive approach, NGPs can gradually shift from being isolated projects to being formally recognized and integrated into national governance frameworks.

Without these foundational steps, integration remains vulnerable to political and administrative turnover as well as donor fatigue. NGPs must evolve from ad hoc pilots into legally anchored systems that redistribute power and resources to IPLCs. While there has been recognition for governments to articulate a clear value proposition for businesses – highlighting the benefits of nature-based action and mitigating operational risks while building stakeholder trust (Arochas and Torres, 2025) – this book argues that equal or greater efforts must be placed on ensuring the value proposition works three ways. This must involve the core group of rights holders – governments and IPLCs – and committed private-sector stakeholders with direct or indirect supply chain interests and investments on the ground. In this formulation, ODA funding, NGO and private sector intermediaries, and foundations will support the core group responsible for negotiating integrated land and resource use plans.

The Integration-Corruption Paradox

In countries like the DRC and Indonesia, integration efforts collide with systemic fragmentation (Prabowo & Cooper, 2016). Ministries of Environment, Agriculture, and Mining often operate as competing fiefdoms (Rainforest Foundation UK, 2025a; Sahide & Giessen, 2015), prioritizing sectoral revenues over cohesive climate action. Chapter 2 presented the current challenge of contested oil sector development in one of the most sensitive ecological systems in the world, the Central Cuvette, which traverses the DRC and the Republic of Congo. For example:

- In the DRC, overlapping jurisdiction between forestry and mining sectors has led to contradictory land-use policies, undermining integrated landscape approaches (Notre Terre Sans Pétrole Coalition, 2025; Sarmiento-Barletti et al., 2022).
- Nearly half of Indonesia's primary forest loss in 2024 – about 242,000 hectares – occurred without an identifiable driver, underscoring gaps in regulatory oversight and permitting, and posing challenges for emissions targets and environmental accountability (Gaveau, 2025). For example, Indonesia's recent energy planning, including large-scale mining permits and expansion in forest zones, directly contradicts its Environment Ministry's emissions targets and climate commitments" (Schlehe, 2024).

Corruption amplifies these divides. Climate finance often flows through intermediaries that capture 40%–60% of revenues (Carbon Market Watch, 2023), while elite networks divert resources meant for communities. Guatemala's UN-backed anti-corruption commission (CICIG) temporarily reduced impunity and enabled forest reforms, but its expiration in 2019 saw elites reassert control, fragmenting climate governance (WOLA, 2019).

Most integration pilots fail to scale because they ignore structural barriers:

1. Donor-Driven Fragmentation: Climate finance flows through disjointed channels, prioritizing isolated metrics over systemic governance. A 2024 UN report found 80% of climate and SDG targets are interdependent, yet funding

mechanisms remain siloed (NDC Partnership, 2024). UNDP recommends that donors move from a siloed approach to a cross-sectoral co-financing model, where costs and benefits are shared among all relevant ministries and sectors in approaching all the SDGs (Remme et al., 2019).
2. Financialization Over Governance: Carbon markets prioritize financial interoperability (e.g., registries, ESG bonds) over governance coherence. In Brazil, the Amazon Soy Moratorium reduced deforestation but failed to align with Indigenous territorial systems, limiting synergistic outcomes (Human Rights Watch, 2019; Nepstad et al., 2019).

Breaking Silos: From Crisis to Coalition Building

Overcoming institutional fragmentation in climate mitigation requires seizing pivotal moments and building unlikely alliances.

The few, some would say rare, examples in Table 6.5 demonstrate how crises, legal precedents, and fiscal pragmatism can catalyze coalitions to break silos and advance integrated governance.

Mechanisms for success in the above examples include:

- Crisis-driven coalitions (e.g., Colombia's post-conflict reforms) create political space for IPLC-state-donor collaboration.
- Grassroots legal advocacy (e.g., FECOFUN's mobilization in Nepal) forces institutional recognition of IPLC rights.
- Donor conditionality (e.g., Norway's \$150M pact with Liberia) aligns finance with tenure security and anti-corruption reforms.

These cases show that silos are most effectively broken when external shocks occur that catalyze cross-sectoral alliances. NGPs can embed crisis-response protocols,

Table 6.5 Silo Breaking Coalitions Are Possible

Trigger	*Key Actors*	*Outcome*
Post-conflict window	IPLC-state-donor coalition	Colombia's participatory land-use planning, reducing deforestation by 34% (NDC Partnership, 2024)
Legal precedent	FECOFUN + reformist judges	Nepal's Forest Act 1993, empowering 22,000 Community Forest User Groups (CFUGs) to manage 34% of national forests (FAO, 2023)
Fiscal pragmatism	Norway's aid agency	Liberia's 2018 Land Rights Act, securing customary tenure amid austerity (World Bank, 2024)

legal empowerment, and adaptive financing as incentives into the terms to be negotiated between the three principal partners.

Integrated Governance: Catalysts for Reform

In climate mitigation, achieving and maintaining legitimacy is a constant challenge. It is built and sustained through ongoing, inclusive engagement, transparent decision-making, and the delivery of tangible results that matter to all stakeholders.

For NGPs to succeed as proposed in this book as fundamental to integrated, appropriate, and feasible governance frameworks that are sustainable – that is, are core to "solutions" – they must foster trust by ensuring that IPLCs, governments, and other actors have a meaningful voice, that processes are fair and accountable, and that outcomes reflect shared, negotiated priorities and pathways forward. Legitimacy grows when systems are open to feedback, adapt to new challenges, and consistently demonstrate both procedural fairness and real-world impact.

The transition from fragmented, siloed climate action to integrated governance rarely occurs organically. It demands seizing pivotal moments and forging unlikely alliances that disrupt entrenched power structures.

The preceding case studies underscore that silos are most effectively broken through crisis-driven coalitions, grassroots legal advocacy, and fiscal pragmatism – principles that must now inform global governance reform. For NGPs to scale, they must institutionalize these lessons, embedding crisis-response protocols, legal empowerment, and iterative learning into their design. The alternative – relying on ad hoc pilots and donor whims – will only perpetuate the fragmentation this chapter critiques.

Pathways for Feasible Integration

To scale integration in weak governance contexts, climate strategies must:

- Embed Hybrid Governance: Blend statutory and customary systems. For example, Namibia's communal conservancies grant communities 90% of tourism and hunting revenues, making conservation more profitable than extraction, with conservancy partnerships creating 1,026 full-time jobs and contributing N$107.5 million to Namibia's economy in 2023 (NACSO, 2024; WWF Namibia, 2023).
- Promote community-controlled management models. For example, while there has been criticism from some around greenwashing of "blue" mangrove carbon offset projects posing as a solution to climate change (Johannessen & Christian, 2023), Colombia's Vida Manglar project reportedly allocates 92% of carbon credit revenue directly to local communities. This allows them to fund conservation, livelihoods, and governance structures without donor-imposed metrics, which is at least a model that reduces dependency on fragmented donor projects by creating self-sustaining financing tied to verifiable outcomes.
- Prioritize Participatory Tools: For example, Kenya's Community Forest Associations used participatory Geographic Information Systems mapping to codify customary land boundaries, reducing elite land grabs while resolving 44 boundary disputes and formalizing community land rights for 31 forests (IFAD, 2023).

Scaling integration requires moving beyond technical fixes to confront power asymmetries. Norway's struggle to break silos – even with strong institutions – highlights that integration is inherently political.

Toward a New Paradigm: Integrated Governance, Local Leadership, and Climate Credibility

If CMLSGS is to succeed, it must move beyond fragmented projects, carbon commodification, and technocratic tools. A viable paradigm must instead be rooted in integrated governance, IPLC leadership, and a clear-eyed assessment of what is politically and socially feasible. The evidence laid out throughout this chapter – indeed, this book – underscores one essential truth: The road to effective mitigation is not paved with financial pledges alone, but with legitimacy and sustained stakeholder negotiation.

Recasting Hybrid Governance as Co-Governance

Hybrid governance offers a framework for blending state, market, and customary systems – but only when anchored in enforceable rights, cultural legitimacy, and IPLC participation. Examples such as Indonesia's FMUs, Colombia's peace accord land titling, and South-South legal literacy exchanges reveal the potential for participatory reform. Models like Akaysha Energy's Social License Framework in Australia and Bolivia's Ley Marco de la Madre Tierra further demonstrate how legal mandates and Indigenous worldviews can reshape finance flows toward community-led stewardship (Māori Climate Commissioner, 2022; Villavicencio-Calzadilla, 2025). Even in longstanding situations where community groups have been empowered in forest management, as in Nepal, the complexity of the management challenges to sustainability remain and must be addressed (Akaysha Energy, 2024; Dhungana et al., 2024; EcoJurisprudence Initiative, 2023).

Addressing Political and Economic Barriers

The most promising models – from Liberia's 2018 Land Rights Law (Wily, 2018) to the LandMark platform's global mapping initiative – succeed because they rebalance power. Legal recognition of customary tenure is foundational. So too is the redirection of climate finance away from intermediaries and toward IPLCs themselves, as demonstrated by the \$1.7 billion Forest Tenure Pledge. Platforms like the Tenure Facility (2022) and IFAD's participatory mapping efforts further show that institutional support and transparent documentation are essential tools for resisting land grabs and sustaining long-term stewardship (IFAD, 2020; RRI, 2020a).

The Risk of Technocratic Overreach

As explored in the preceding sections, technocratic tools and scientific urgency have become dominant forces in shaping climate action. But when urgency trumps feasibility – or when "the science" is invoked to suppress negotiation – the result is often unsustainable interventions that ignore social context. Initiatives like the Carbon Markets Initiative and Integrated Climate Action Planning offer technical promise but must be grounded in real-world legitimacy and transparency (C40 Cities, 2022; RMI, 2023). Likewise, science-based approaches like FLR, PES, and CBFM will only succeed if they account for power dynamics, elite capture, and the lived experiences of IPLCs (Chomba et al., 2021; Ostrom, 1990; Pagiola et al., 2014).

The Role of ESG Actors: From Compliance to Commitment

Corporate actors and ESG investors must go beyond reputational signaling to co-finance and co-govern sustainable land management. Whether through Safaricom's potential advocacy for the Community Land Act in Kenya, ExxonMobil's participatory mapping in PNG (ExxonMobil PNG Limited., 2016), BlackRock's leverage over deforestation-free supply chains in Brazil, or financial institutions like Rabobank – whose forest-risk lending in Brazil has been linked to billions in social and environmental damages (Rijk et al., 2023) – the tools already exist. What's needed is the political will and regulatory framework to align ESG practices with long-term IPLC empowerment, not just market returns (Ford Foundation, 2021; LandMark, 2024). This will be fundamental to any future AFOLU success where co-benefits are central to success (Bustamente et al, 2014).

Developing New Metrics

A persistent weakness of mainstream climate mitigation frameworks is their reliance on externally defined, top-down metrics, typically focused on carbon, donor disbursement, or project outputs, that rarely reflect the priorities or realities of IPLCs (Boltz et al., 2022; CARE, 2015). To achieve legitimate and durable outcomes, new metrics must not be imposed but negotiated through the NGP platforms themselves. These negotiated metrics become the operational expression of shared priorities, trade-offs, and locally defined success, ensuring that measurement, finance, and accountability are rooted in IPLC perspectives and adaptive to context. See Table 6.6.

Table 6.6 From Less Relevant to More Relevant Negotiated Metrics Old Metrics Negotiated Metrics (via NGPs)

• Carbon tons reduced/sequestered (CO_2e)	• Multi-criteria outcomes: carbon, biodiversity, land tenure, livelihoods, and culture (all defined through negotiation)
• Donor disbursement rates	• Milestones and indicators set by the NGP consensus, reflecting both IPLC and other stakeholder priorities
• Project outputs (e.g., trees planted; hectares under "improved management")	• Locally relevant social, ecological, and cultural well-being indicators are key to sustainability.
• Top-down reporting requirements	• Participatory, community-owned monitoring and evaluation, agreed upon by NGPs in IPLC-managed landscapes
• Short-term project cycles	• Long-term, adaptive assessment cycles, revisited through NGP negotiation
• Externally defined indicators	• Indicators co-created and adaptively revised through ongoing NGP deliberation

Mobilizing NGPs Through Learning and Adaptive Pilots

NGPs and the integration of stakeholders in complex planning processes will ultimately succeed once governments, the private sector, and IPLCs see common interest. Other stakeholders, including BINGOs, research institutions, and, of course, the gamut of donors, have the opportunity *now* to pivot from a top-down planning modus operandi to actual support of bottom-up, collaborative negotiated processes at country and jurisdictional levels. As noted, some like TNC are taking steps by prioritizing and establishing more effective and genuine partnerships with IPLCs (The Nature Conservancy, 2022). Evaluating the impact of this trend will be important. This also builds on what USAID referred to as localization (2022).

As with any project that needs iterative testing and refinement before scaling, based on evidence and local feedback (van Noordwijk et al., 2001), NGPs will be no different. The learning challenges for NGPs will include the following:

- Design and implementation of adaptive pilot projects that test integration at multiple scales and in diverse contexts, with robust mechanisms for learning from failure, not just success (Catalano et al., 2018).
- Facilitation of peer learning and South-South exchange among IPLCs and government actors to share operational lessons and build political momentum for reform is actively supported by UN DESA through multistakeholder forums and knowledge exchange initiatives (UN DESA, 2024).
- Institutionalization of iterative review processes that allow for course correction and adaptation as contexts and priorities evolve (van Noordwijk et al., 2001).

Ironically, the fiscal challenges that the conservation and development community now faces with CMLSGS also offer the opportunity for a long-overdue paradigm shift, which will only be realized if and when the three key stakeholder groups – government, IPLCs, and the gamut of private sector operators impacting the Global South – perceive common ground and seize the moment.

Conclusion: Toward Negotiated Governance

The lessons are clear. Integration cannot remain rhetorical. It must be built into laws, institutions, and accountability mechanisms that endure beyond project cycles. IPLC tenure security is not a supplement to climate mitigation; it is a precondition as both an ongoing process of maintaining security and an outcome enabling investment. Yet, as Ribot and Peluso (2003) emphasize, securing legal rights alone is insufficient unless IPLCs have genuine access to land, resources, and benefits, which is shaped by broader power dynamics and institutional arrangements.

Participation, meanwhile, cannot be performative; it must be enabled, and its minimum or maximum threshold must be negotiated. Governance must not be technocratic; it must be credibly inclusive to lead to sustainability of results and outcomes. And science, rather than being used as a cudgel for urgency, must serve as a tool for clarifying trade-offs and enabling consensus to be reached. Such consensus can not only consider purported financial returns to external stakeholders while promising safeguards, primarily, to IPLCs. Why would IPLCs agree to cede rights for meager hypothetical returns in years in countries where corruption is high and a lack of trust in government, the private sector (and frequently even BINGOs) is high?

Only by grounding climate mitigation in legitimate, place-based governance can the promises of carbon finance, biodiversity conservation, and sustainable development be realized in the Global South. This involves figuring out in specific contexts how institutional silos that serve as barriers for cross-sectoral collaboration can be broken down.

The path forward lies in establishing NGPs, socially licensed finance, and a redefinition of what success looks like, one that centers IPLC priorities, platforms that enable "broad processes" and negotiation of cross-sectoral barriers or silos (Oseland, 2019), not just carbon metrics.

To overcome these challenges to generate synergies and break down silos, steps can be taken:

1. **Capacity Building**: Training programs can empower IPLCs with negotiation skills and knowledge about their legal rights. For instance, IFAD has supported participatory mapping initiatives that help IPLCs document their territories effectively while fostering local governance systems (IFAD, 2020). Weak implementation capacity in both state institutions and IPLCs represents a significant challenge to moving from isolated pilots to wider implementation. Building local governance capacity through demand-driven training programs, legal literacy initiatives, and adaptive management strategies will be essential to ensure NGPs function effectively.
2. **Formal Tenure Security or "Secure-Enough" Tenure Moving Forward?:** Answering this question is a key integration challenge, yet one that is far from straightforward. The seemingly favorable case for mobilizing customary land for development in Papua New Guinea illustrates that "the incompatibility of customary social institutions with the operation of a capitalist economy that includes a land market" (Filer, 2019) is not readily overcome, echoing Nuhu's concern (2021) about applying land access governance to overcome blockages in the land sector. This dilemma is not simply institutional but ontological: as Weiner and Glaskin (2007) show, efforts to codify or commodify "customary" tenure transform the very relationships they aim to formalize, producing hybrid arrangements in PNG that struggle to deliver either cultural legitimacy or full market efficiency.
3. **Financial Support**: Innovative funding mechanisms, such as direct grants or simplified financing models, can help IPLCs overcome economic barriers. The $1.7 billion forest tenure pledge is an example of global recognition of this need (Butler, 2021).
4. **Legal Advocacy**: Strengthening advocacy for IPLC rights through international frameworks like FPIC ensures their voices are heard in policy decisions affecting their lands (United Nations, 2007).
5. **Tailored strategies**: Mobilizing broad processes, political will, and institutional entrepreneurs (Oseland, 2019) is true whether in Norway or the DRC.

Framing Integration as a Negotiated Process

As this chapter has shown, "integration" is too often deployed as a policy catchword without a clear operational meaning. To make integration actionable in the context of CMLSGS, it must be understood as a multidimensional and negotiated governance process. Drawing on Reed, Ros-Tonen, and Sunderland (Reed et al., 2020), five core components define effective integration:

1. Integration across sectors – Overcoming silos between forestry, agriculture, conservation, energy, and development to enable coherent planning across ministries and mandates.
2. Integration across scales – Ensuring that decision-making aligns across community, district, provincial, national, and international levels.
3. Social integration – Engaging IPLCs, civil society, local governments, private sector actors, and donors in iterative dialogue and joint planning.
4. Integration of objectives and interests – Negotiating trade-offs and synergies across climate, biodiversity, livelihood, and cultural priorities rather than assuming they align.
5. Integration of knowledge systems – Weaving together scientific, traditional, and experiential knowledge in ways that support locally grounded adaptive management.

For integration to be effectively operationalized, each stakeholder group – governments, IPLCs, the private sector, and civil society – must have clearly defined roles and accountability mechanisms. Governments should provide legal and policy frameworks that support cross-sectoral governance, including the integration of climate, agriculture, and biodiversity ministries. Private sector actors must align their ESG goals with IPLC rights and land stewardship practices, ensuring that economic incentives for deforestation-free supply chains are not just financial but also deeply aligned with the priorities of local communities. Civil society and NGOs should act as facilitators and monitors, ensuring that IPLC participation is not tokenistic but active and influential in decision-making. Integration must move beyond mere coordination and require co-design at every level – from local planning to national policy – ensuring that all sectors are equally represented in decision-making processes.

These elements underscore that integration is not a blueprint but a process. As Reed et al. emphasize, "integrated landscape approaches are inherently context specific, and their design and implementation require negotiation among stakeholders to balance differing objectives, manage unavoidable trade-offs, and build consensus for action." This aligns directly with the paradigm shift advanced throughout this book: Negotiated governance must replace top-down programming, primarily driven by carbon mitigation logic to the detriment of feasibility.

While concepts like polycentric governance (Ostrom, 2009), institutional hybridity (Ribot & Larson, 2012), and integrated landscape approaches (Reed et al., 2020) offer valuable frameworks for inclusive and adaptive climate governance, they often fall short of proposing operational structures that can be embedded in national systems or scaled across diverse political economies. This book builds on that foundation but goes further. It proposes tripartite national and subnational negotiation platforms as a practical, scalable mechanism for aligning governance with IPLC realities, political feasibility, and the demands of climate legitimacy.

These platforms are not merely consultative spaces – they are structured co-governance arrangements involving IPLCs, governments, and the private sector. They enable transparent negotiation of consent, benefit-sharing, land-use planning, and grievance mechanisms. Importantly, they emphasize IPLC capacity development and are designed to resolve trade-offs between carbon mitigation, livelihoods, and tenure security through context-specific, negotiated processes.

Table 6.7 compares this model with leading concepts from the literature to clarify how it extends current thinking and operationalizes a deeper paradigm shift.

Table 6.7 What Makes Negotiation Governance Platforms Different?

Governance Concept	*Key Characteristics*	*Limitations*	*How Tripartite Negotiation Platforms Extend It*
Polycentric Governance (Ostrom, 2009)	Multiple, overlapping centers of decision-making; local autonomy; subsidiarity; cooperation and learning across levels	Often lacks mechanisms for resolving power asymmetries or coordinating actors with vastly unequal resources	Adds formal structure to polycentricity by requiring co-decision among IPLCs, government, and private sector actors within a single platform with clear mandates
Institutional Hybridity (Ribot & Larson, 2012)	Coexistence and interaction between state institutions and customary/local governance systems	Can result in ad hoc or unstable arrangements without mechanisms for dispute resolution or scaling	Formalizes hybridity into co-governance with legal recognition, procedural clarity, and accountability – moving from coexistence to co-creation
Negotiated Governance/ Integrated Landscape Approaches (Reed et al., 2020)	Integration is negotiated across sectors, stakeholders, objectives, and knowledge systems; emphasizes trade-off management	Conceptually rich but often not institutionalized in climate finance frameworks or policy architecture	Operationalizes negotiation through platforms with built-in STAM alignment, consent, and shared legitimacy – moving from process to policy
Negotiated Governance Platforms (*This book's contribution*)	Structured co-governance mechanism between IPLCs, governments, and private/philanthropic sectors; centers IPLC rights, FPIC, tenure security, and capacity development	Still experimental and politically ambitious; requires donor, state, and corporate buy-in	Offers a replicable governance model for scaling integrated, IPLC-led climate action – grounded in appropriateness, feasibility, and negotiated legitimacy

The political feasibility of such integration also depends on shifting from siloed ministries and donor mandates that reinforce power asymmetries to platforms where adaptive co-design, joint decision-making, cross-sectoral coordination, and negotiated governance are incentivized and institutionalized (Catalano et al., 2019; Lah, 2025; Reed et al., 2020). Without structural mechanisms to align competing interests and power asymmetries, even well-designed integration efforts will remain constrained by fragmentation and inertia.

Institutionalizing Integration: From Pilots to Durable Governance – A Realistic View

While innovative models like NGPs and negotiated metrics are essential, they will not deliver sustained impact unless embedded in robust, multi-level governance frameworks. Experience across Africa, Asia, and the MENA region shows that weak decentralization, unclear mandates, and lack of cross-sectoral legal frameworks routinely undermine climate action, even when local pilots succeed (see Ampaire et al., 2020; IPCC AR6 WG3, Chapter 13: National and sub-national policies and institutions, 2022). To move from rhetoric to reality, governments must enact legal and regulatory reforms, define clear roles and resource flows, embed negotiated agreements in statutory law, and build durable accountability structures (see Rights and Resources Initiative, 2023; UN DESA, 2023). So too, synergies between mitigation and adaptation must be created.

In practice, where corruption is entrenched and power is exercised opaquely or with force, even the most reasoned and appealing calls for reforms can founder. As discussed in Chapter 2, the campaign against oil block development in the DRC's Cuvette Centrale, led by the Notre Terre Sans Pétrole coalition, shows both the potential and the limits of civil society advocacy. This coalition of over 170 Congolese and international organizations has mobilized mass petitions, legal challenges, transparency campaigns, and international solidarity to oppose oil extraction in one of the world's most important peatland ecosystems, harboring at least 30 billion metric tons of carbon in both DRC and RoC's Cuvette Central (also see BankTrack et al., 2024; Rainforest Foundation UK, 2025; Rijk, 2023). Despite these efforts, the DRC government has pressed ahead, reopening bidding for 52 new oil blocks and making clear that civil society alone cannot reverse deeply embedded extractive interests.

The main lesson is not that advocacy or new platforms guarantee success, but that civil society action remains essential for:

- Raising the political and reputational costs of corrupt or destructive decisions;
- Delaying or disrupting harmful projects, even if only temporarily;
- Exposing legal and procedural flaws and keeping the spotlight on government and corporate accountability (see Global Witness, 2018);
- Building solidarity, capacity, and alternative visions for the future, so that when opportunities for reform do arise, there is a foundation to build on (see Rainforest Rescue, 2025).

In high-corruption, high-impunity settings, history shows that most "win-win" development models are vulnerable to elite capture, bureaucratic inertia, and project failure (see Dick-Sagoe et al., 2023). Civil society's role is not to guarantee transformation but to resist the worst outcomes, empower communities, and keep alive the possibility of more accountable, negotiated governance, whether in contexts where transnational corruption is a factor or where domestic policies and practices are in question (see Maslen, 2024). This is a precondition for establishing enabling conditions to achieve synergies in climate mitigation and adaptation as governments aspire to.

In Sum

Civil society advocacy, like that of Notre Terre Sans Pétrole in the Cuvette Centrale, cannot by itself overcome systemic corruption in the DRC or elsewhere (Transparency International, 2024). Nor can advocacy alone guarantee effective integration. But civil society advocacy is a necessary force for accountability, transparency, and hope, even when progress is incremental or uncertain. In the real world, this is not about "transformation" on paper. Rather, it is about refusing to cede the future to corruption related to a lack of transparency in deciding resource management decisions implicating fragile ecosystems and hundreds of thousands of the world's poorest people. This simply perpetuates the infamous "resource curse" dynamic that has plagued the Congo and other countries of the Global South for hundreds of years.

References

Agrawal, A., & Ostrom, E. (2001). Collective action, property rights, and decentralization in resource use in India and Nepal. *Politics & Society*, 29(4), 485–514.

Akaysha Energy. (2024). Social License Framework. Retrieved from https://akayshaenergy.com/about/social-license-framework

Altamirano, J. C., Finch, M., & Akhtar, S. (2022). Sink or swim: How Indigenous and community lands can make or break nationally determined contributions. Forest Declaration Assessment Briefing Paper. Retrieved from https://forestdeclaration.org/wp-content/uploads/2022/03/Sink-or-swim-IPLC-lands-and-NDCs.pdf

Amnesty International. (2023). Human rights in Papua New Guinea. Retrieved from https://www.amnesty.org/en/location/asia-and-the-pacific/south-east-asia-and-the-pacific/papua-new-guinea/report-papua-new-guinea/

Ampaire, E. L., Jassogne, L., Providence, H., Acosta, M., Twyman, J., Winowiecki, L., & van Asten, P. (2017). Institutional challenges to climate change adaptation: A case study on policy action gaps in Uganda. In Environmental Science & Policy (Vol. 75, pp. 81–90). Elsevier BV. https://doi.org/10.1016/j.envsci.2017.05.013

Anaya, J. (2009). *Report of the Special Rapporteur on the Situation of Human Rights and Fundamental Freedoms of Indigenous People*. United Nations Human Rights Council.

Angelsen, A., Martius, C., De Sy, V., Duchelle, A. E., Larson, A. M., & Pham, T. T. (2018). *Transforming REDD+: Lessons and New Directions*. CIFOR.

Arochas, A. & Torres, C. (2025). Chile shows collaboration builds nature-positive economies. *World Economic Forum*. https://www.businessfornature.org/news/chile-collaboration-nature-economies

Bäckstrand, K. (2006). Multi-stakeholder partnerships for sustainable development: Rethinking legitimacy, accountability and effectiveness. *European Environment*, 16(5), 290–306.

BankTrack, 350.org Africa, Greenpeace Africa, & Rainforest Foundation UK. (2024, October 18). *DRC Oil & Gas Auction*. BankTrack. BankTrack – DRC Oil & Gas Auction. https://www.banktrack.org/project/democratic_republic_of_congo_oil_gas_auction

Barrage, L., & Nordhaus, W. D. (2024). Policies, projections, and the social cost of carbon: Results from the DICE-2023 model. *Proceedings of the National Academy of Sciences*, 121(13), e2312030121. https://doi.org/10.1073/pnas.2312030121

Bilal, U., & Känzig, D. (2024). The macroeconomic impact of climate change: Global vs. local. *Working Paper No. 24-5*. Federal Reserve Bank of Boston. Accessed at: https://www.lse.ac.uk/CFM/assets/pdf/BilalA-MacroeconomicImpact.pdf

Bimir, M. N., & Matus, K. J. M. (2025). From silos to joined-up government? Examining national climate policy integration in Ethiopia. *Climate Policy*, Advance online publication, March 2025. https://doi.org/10.1080/14693062.2025.2479809

Boltz, F., Losos, E., Karasik, R., & Mason, S. (2022). *Developing key performance indicators for climate change adaptation and resilience planning*. Nicholas Institute for Environmental Policy Solutions, Duke University. https://nicholasinstitute.duke.edu/publications/developing-key-performance-indicators-climate-change-adaptation-and-resilience

Bonis-Charancle, J. M., Tiani, A. M., Brown, M., Neba, G. A., Mogba, Z., Lescuyer, G., Warne, R., & Greenberg, B. (2009). Strengthening local analytical capability: Community options analysis and investment. In M. C. Diaw, T. Aseh, & R. Prabhu (Eds.), *In Search of Common Ground: Adaptive Collaborative Management in Cameroon* (pp. 275–300). Bogor, Indonesia: Center for International Forestry Research (CIFOR). https://www.cifor-icraf.org/knowledge/publication/2744/

Brown, M. I. (2013). *Redeeming REDD: Policies, Incentives, and Social Feasibility for Avoided Deforestation*. Routledge.

Brown, D., & MacLellan, M. (2020). A Multiscalar and Justice-Led Analysis of REDD+: A Case Study of the Norwegian–Ethiopian Partnership. *Global Environmental Politics*, 20(1), 11–37. https://doi.org/10.1162/glep_a_00536

Bustamante, M., Robledo-Abad, C., Harper, R., Mbow, C., Ravindranat, N. H., Sperling, F., . . . & Smith, P. (2014). Co-benefits, trade-offs, barriers and policies for greenhouse gas mitigation in the agriculture, forestry and other land use (AFOLU) sector. *Global Change Biology*, 20(10), 3270–3290.

Carbon Market Watch. (2023). Secretive Intermediaries: Are carbon markets really financing climate action? https://carbonmarketwatch.org/wp-content/uploads/2023/02/CMW-briefing-on-intermediaries-1.pdf

CARE (2015). *Framework of Milestones and Indicators for Community-Based Adaptation*. Geneva: CARE International.

Catalano, A. S., Redford, K., Margoluis, R., & Knight, A. T. (2018). Black swans, cognition, and the power of learning from failure. *Conservation Biology*, 32, 584–596. https://conbio.onlinelibrary.wiley.com/doi/full/10.1111/cobi.13045

Chazdon, R. L., Brancalion, P. H., Laestadius, L., Bennett-Curry, A., Buckingham, K., Kumar, C., . . . & Wilson, S. J. (2017). When is a forest a forest? Forest concepts and definitions in the era of forest and landscape restoration. *Ambio*, 46(6), 557–569.

Chomba, S., Mwangi, A. G., & Karanja, F. K. (2021). Forest landscape restoration in Kenya: Progress, challenges, and policy implications. *Forest Policy and Economics*, 129, 102506.

ClientEarth. (2023). Community-Based Forest Management: Legal Frameworks in Five Congo Basin Countries. https://www.clientearth.org/latest/documents/community-based-forest-management-legal-frameworks-in-five-congo-basin-countries/

Climate Horizons. (2024). *Mapping and Trends Analysis on Just Transition Initiatives*. Laudes Foundation, Wallace Global Fund, Ford Foundation.

Climate Investment Funds (CIF). (2024). Bolder Ambitions, More Urgency, Unflinching Determination: Annual Report 2023. Climate Investment Funds (CIF). Available at: https://www.cif.org/sites/cif_enc/files/knowledge-documents/cif_annual_report-23.pdf

Climate Policy Initiative. (2023). Global landscape of climate finance 2023. https://www.climatepolicyinitiative.org/publication/global-landscape-of-climate-finance-2023/

Conservation International. (2022). Indigenous negotiations resource guide. *Conservation International.* https://www.conservation.org/projects/indigenous-negotiations-resource-guide

Crawford, A. & Church, C. (2019). *Engaging the private sector in National Adaptation Planning Processes.* Winnipeg, Canada: International Institute for Sustainable Development. Retrieved from www.napglobalnetwork.org

Cronkleton, P., Taylor, P. L., Barry, D., Stone-Jovicich, S., & Schmink, M. (2011). Environmental governance and the emergence of forest-based social movements. *CIFOR Occasional Paper 66.*

Cultural Survival. (2022, June 26). Indigenous peoples continue to push for rights, inclusion, and direct financing in climate change solutions. *Cultural Survival News.* https://www.culturalsurvival.org/news/indigenous-peoples-continue-push-rights-inclusion-and-direct-financing-climate-change?gad_source=1&gad_campaignid=144221-9742&gbraid=0AAAAAD9zhqy9XvzQEzCeKcwDvbR1BdLF5&gclid=Cj0KCQjw6bfHBhDNARIsAIGsqLiDWCgw1wcYF8j2ST13veFcH70HkjhL0ymAJi1ESXRWnD8k_UVV4pMaApYVEALw_wcB

Dahal, Ganga Ram & Chapagain, Apsara. (2005). "Community forestry in Nepal: decentralized forest governance." CIFOR. https://www.cifor-icraf.org/publications/pdf_files/events/documentations/yogyakarta/papers/chapter%205%20dahal.pdf

Deininger, Franziska; Woodhouse, Andrea; Kuriakose, Anne T.; Gren, Ana; Liaqat, Sundas. (2023). Placing Gender Equality at the Center of Climate Action. World Bank Group Gender Thematic Policy Notes Series;Issues and Practice Note. © World Bank. http://hdl.handle.net/10986/39436 License: CC BY 3.0 IGO.

Del Río Duque, M. L., Rodríguez, T., Pérez Lora, Á. P., Löhr, K., Romero, M., Castro-Nunez, A., Sieber, S., & Bonatti, M. (2022). Understanding systemic land use dynamics in conflict-affected territories: The cases of Cesar and Caquetá, Colombia. *PLOS ONE,* 17(5), e0269088. https://doi.org/10.1371/journal.pone.0269088

Devex. (2025). Trump budget proposes unprecedented, "reckless" cuts to foreign aid. https://www.devex.com/news/trump-budget-proposes-unprecedented-reckless-cuts-to-foreign-aid-109988

Dhungana, N., C.-H. Lee, C. Khadka, S. Adhikari, N. Pudasaini, & P. Ghimire. 2024. "Evaluating Community Forest User Groups (CFUGs) Performance in Managing Community Forests: A Case Study in Central Nepal." *Sustainability* 16: 4471.

Dick-Sagoe, C., Lee, K.Y., Odoom, D. et al. (2023). Stakeholder perceptions on causes and effects of public project failures in Ghana. *Humanities and Social Sciences Communications,* 10, 14. https://doi.org/10.1057/s41599-022-01497-7

Dooley, K., Keith, H., Larson, A., Catacora-Vargas, G., Carton, W., Christiansen, K. L., Enokenwa Baa, O., Frechette, A., Hugh, S., Ivetic, N., Lim, L. C., Lund, J. F., Luqman, M., Mackey, B., Monterroso, I., Ojha, H., Perfecto, I., Riamit, K., Robiou du Pont, Y., & Young, V., 2022. The Land Gap Report 2022. Available at: https://www.landgap.org

Doyle, C., & Cariño, J. (2013). *Making Free, Prior & Informed Consent a Reality: Indigenous Peoples and the Extractive Sector.* Middlesex University School of Law and Indigenous Peoples Links.

Duchelle, A. E., Simonet, G., Sunderlin, W. D., & Wunder, S. (2018). What is REDD+ achieving on the ground? *Current Opinion in Environmental Sustainability,* 32, 134–140. https://www.sciencedirect.com/science/article/pii/S1877343517301872

Duchelle, A. E., et al. (2023). Reassessing REDD+ outcomes: A decade of implementation. *Environmental Research Letters,* 18(6), 064002.

EcoJurisprudence Initiative. (2023). Bolivia Law No. 300 Framework Law of Mother Earth and Integral Development to Live Well. https://ecojurisprudence.org/initiatives/law-no-300-the-framework-law-of-mother-earth-and-integral-development-to-live-well/

Ecosystem Marketplace. (2025, April 10). *Nature's Investment Frontier: Practical Paths Forward for Biodiversity Markets and Finance.* Forest Trends. https://www.ecosystemmarketplace.com/wp-content/uploads/2025/04/Natures-Investment-Frontier.pdf

European Commission. (2023). Special eurobarometer 538: Climate change. Directorate-General for Communication. https://europa.eu/eurobarometer/surveys/detail/2954

ExxonMobil PNG Limited. (2016). Land Access, Resettlement and Livelihood Restoration Management Plan – Production Phase. *PNG LNG Project Environmental and Social Management Plan*. Retrieved from https://www.pnglng.com/media/PNG-LNG-Media/Files/Environment/Production%20ESMP/Land-Access-Resettlement-and-Livelihood-Restoration-Management-Plan-Rev-1_Nov-2016.pdf

Fairbrother, M. (2022). Public opinion about climate policies: A review and call for more studies of what people want. *PLOS Climate*, 1(3), e0000030. https://doi.org/10.1371/journal.pclm.0000030

FAO. (2023). *Transforming the Future of Community Forestry in Nepal*. Forest and Farm Facility. Food and Agriculture Organization of the United Nations.

Ferraro, P. J., & Hanauer, M. M. (2014). Through what mechanisms do protected areas affect environmental and social outcomes? *Philosophical Transactions of the Royal Society B*, 369(1653), 20130345.

Filer, C. (2019). Two steps forward, two steps back: The mobilisation of customary land in Papua New Guinea. *Development Policy Centre Discussion Paper No. 86*. Crawford School of Public Policy, Australian National University, Canberra. https://devpolicy.org/publications/discussion_papers/DP86-Mobilisation-of-customary-land-in-Papua-New-Guinea-December-2019.pdf

Financial Times. (2025, April 25). "ESG fund outflows hit record as sustainable investing backlash grows." https://www.ft.com/content/9f425c25-4fc3-45de-bcc5-e9c75d6d14d3

Ford Foundation. (2021). COP26 forest tenure pledge. https://www.fordfoundation.org/news-and-stories/stories/cop26-forest-tenure-pledge/

Ford Foundation. (2025). How Indigenous Peoples and Local Communities Are Shaping the Future of Climate Funding. Retrieved from https://www.fordfoundation.org/news-and-stories/stories/how-indigenous-peoples-and-local-communities-are-shaping-the-future-of-climate-funding/

Forest Declaration Assessment. (2024). *Forest Finance Flows: Progress Toward 2030 Goals*. Forest Declaration Platform. https://forestdeclaration.org/wp-content/uploads/2024/10/2024ForestDeclarationAssessment.pdf

Forest Peoples Programme (FPP). (2024, March 3). *New Bahasa Indonesia Community Guide to Forest Stewardship Council's Remedy Framework*. Forest Peoples Programme. https://www.forestpeoples.org/publications-resources/reports/article/new-bahasa-indonesia-community-guide-to-forest-stewardship-councils-remedy-framework/

Freedom House. (2024). Freedom in the World 2024: The global expansion of authoritarian influence. Retrieved from https://freedomhouse.org/sites/default/files/2024-02/FIW_2024_DigitalBooklet.pdf

FT Longitude. (2025, February 6). The DEI paradox: Navigating backlash and progress in 2025. Retrieved from https://longitude.ft.com/blog/the-dei-paradox-navigating-backlash-and-progress-in-2025/

Gallup. (2024). Confidence in institutions poll. https://news.gallup.com/poll/1597/confidence-institutions.aspx

Garnett, S. T., Burgess, N. D., Fa, J. E., Fernández-Llamazares, Á., Molnár, Z., Robinson, C. J., . . . & Leiper, I. (2018). A spatial overview of the global importance of Indigenous lands for conservation. *Nature Sustainability*, 1(7), 369–374. https://www.nature.com/articles/s41893-018-0100-6

Gaveau, D. (2025, February 24). Indonesia's Primary Forest Loss slows in 2024. Nusantara Atlas. https://nusantara-atlas.org/indonesias-primary-forest-loss-slows-in-2024/

Ghimire, H. R. (2024). *Enhancing Internal Governance Within a National Federation of Community Forestry Users in Nepal: The Case of FECOFUN*. London: International Institute for Environment and Development (IIED). https://www.iied.org/22351g

Global Dialogue. (2024). Summary Report of the Global Dialogue: Enabling Indigenous Peoples and Local Communities to Engage in and Benefit from Results-Based Climate

Finance, San José, Costa Rica, November 6–9, 2023. Forest Carbon Partnership Facility & EnABLE Fund, World Bank. https://www.forestcarbonpartnership.org/sites/default/files/2024-05/global_dialogue_summary_report.pdf
Global Witness. (2018, July 24). Total systems failure: Exposing the global secrecy destroying forests. https://globalwitness.org/en/campaigns/forests/total-systems-failure
Green Climate Fund. (2022). Indigenous Peoples Policy: Implementation progress report. *Green Climate Fund Board Document.*
Greenpeace Africa. (2023, September 8). Blood oil: The threat from DRC oil auctions in the Cuvette Centrale peatlands. c9d5ab91-blood-oil-gpa-report-aug-2023-v5.pdf
Harvey, C., & Schonhardt, S. (2024, September 25). Carbon markets take center stage at climate week as criticisms abound. *E&E News.*
Herrera-Franco, G., Bollmann, H. A., Lofhagen, J. C. P., Bravo-Montero, L., & Carrión-Mero, P. (2023). Approach on water-energy-food (WEF) nexus and climate change: A tool in decision-making processes. *Environmental Development, 46,* 100858. https://www.researchgate.net/publication/369905363_Approach_on_water-energy-food_WEF_nexus_and_climate_change_A_tool_in_decision-making_processes
Hulme, M. (2009). *Why we Disagree about Climate Change: Understanding Controversy, Inaction and Opportunity.* Cambridge University Press.
Human Rights Watch. (2019). Rainforest Mafias: How violence and impunity fuel deforestation in Brazil's Amazon. Retrieved from https://www.hrw.org/report/2019/09/17/rainforest-mafias/how-violence-and-impunity-fuel-deforestation-brazils-amazon
Hunsberger, C., Corbera, E., & Borras, S. M. (2017). Climate change mitigation, land grabbing and conflict: Towards a landscape-based and collaborative action research agenda. *Canadian Journal of Development Studies/Revue canadienne d'études du développement*, 38(3), 305–322. https://doi.org/10.1080/02255189.2016.1250617
IEA. (2025). *Electricity 2025.* IEA, Paris. https://www.iea.org/reports/electricity-2025
Inclusive Conservation Initiative. (2022). *Inclusive Conservation Initiative: Project Document.* IUCN & Conservation International/Global Environment Facility.
International Fund for Agricultural Development (IFAD). (2009). *Good Practices in Participatory Mapping.* Rome: IFAD. https://www.ifad.org/documents/38714170/39144386/PM_web.pdf/7c1eda69-8205-4c31-8912-3c25d6f90055
International Fund for Agricultural Development (IFAD). (2020). The Land Tenure Security Advantage: A catalytic asset for sustainable and inclusive rural transformation. Rome: IFAD. https://www.ifad.org/documents/38714170/42014188/landtenure_security_advantage.pdf
International Institute for Sustainable Development (IISD). (2021). *Addressing Climate Change through Integrated Responses: Linking Adaptation and Mitigation.* https://www.iisd.org/publications/reports/addressing-climate-change-linking-adaptation-mitigation
International Monetary Fund (IMF). (2022, October). Scaling up private climate finance in emerging market and developing economies: Challenges and opportunities. In *Global Financial Stability Report*, Chapter 2. Washington, DC: IMF. https://www.imf.org/-/media/Files/Publications/GFSR/2022/October/English/ch2.ashx
Jansen, B. (2025, May 2). Trump budget proposal aims to cut $163 billion, mirroring DOGE plan. *USA Today.* Retrieved from https://www.usatoday.com/story/news/politics/2025/05/02/trump-budget-163-billion-cuts/83405569007/
Johannessen, S. C., & Christian, J. R. (2023). Why blue carbon cannot truly offset fossil fuel emissions. *Communications Earth & Environment*, 4, 411. https://www.nature.com/articles/s43247-023-01068-x
Kassa, H., Foli, S., Kanshie, A., & Kassa, B. (2020). Forest landscape restoration and its implications for ecosystem services and resilience in Ethiopia. In *Nature-Based Solutions for Resilient Ecosystems and Societies* (pp. 353–372). Springer.
Kennedy, B., & Tyson, A. (2024, December 9). How Americans view climate change and policies to address the issue: Mixed views of how climate policies impact the U.S. economy. *Pew Research Center.* https://www.pewresearch.org/science/2024/12/09/how-americans-view-climate-change-and-policies-to-address-the-issue/

Kroc Institute for International Peace Studies. (2024, May 8). Eighth Report on Colombian Peace Agreement Implementation Released. Notre Dame News. Jena O'Brien. Retrieved from https://news.nd.edu/news/authors/jena-obrien/

Lah, O. Breaking the silos: integrated approaches to foster sustainable development and climate action. *Sustain Earth Reviews* 8, 1 (2025). https://sustainableearthreviews.biomedcentral.com/articles/10.1186/s42055-024-00102-w

LandMark. (2024). Global platform of Indigenous and community lands. https://www.landmarkmap.org

Lang, C. (2024, November 12). COP29 approval of Article 6.4 is "opening the floodgates for a global carbon market that will have devastating impacts on communities in the Global South." *REDD-Monitor.* https://reddmonitor.substack.com/p/cop29-approval-of-article-64-is-opening

Lang, C. (2025, July 3). Carbon markets are a failed model of climate finance that prioritises profits over people and delays over action. *REDD-Monitor*. Retrieved from: https://reddmonitor.substack.com/p/carbon-markets-are-a-failed-model

Laudes Foundation, Wallace Global Fund, & Ford Foundation. (2024, September 19). *Mapping and Trend Analysis on Just Transition Initiatives*. Climate Horizons for Laudes Foundation.https://www.laudesfoundation.org/latest/news-and-stories/stories/2024/mapping-and-trend-analysis-on-just-transition-initiatives/

Locatelli, B., Pavageau, C., Pramova, E., & Di Gregorio, M. (2015). Integrating climate change mitigation and adaptation in agriculture and forestry: Opportunities and trade-offs. *Wiley Interdisciplinary Reviews: Climate Change*, 6(6), 585–598. https://wires.onlinelibrary.wiley.com/doi/10.1002/wcc.357

Lomborg, B. (2021). *False Alarm: How Climate Change Panic Costs Us Trillions, Hurts the Poor, and Fails to Fix the Planet*. Hachette Books.

Lowy Institute. (2025, February 25). UK aid cut: Implications for an increasingly lonely Australia. *The Interpreter*. Retrieved from https://www.lowyinstitute.org/the-interpreter/uk-aid-cut-implications-increasingly-lonely-australia

Lund, J. F., Sungusia, E., Mabele, M. B., & Scheba, A. (2017). Promising change, delivering continuity: REDD+ as conservation fad. *World Development*, 89, 124–139. https://www.sciencedirect.com/science/article/abs/pii/S0305750X15312821

Luttrell, C., and Pellini, A. (2025). Lessons from MRLG for the design of land governance programmes. Brief. Capitalisation of Experience Series #5. Vientiane: MRLG.

Maslen, C. (2024). *The Role of Civil Society Organisations in Countering Transnational Corruption*. Bergen: U4 Anti-Corruption Resource Centre, U4 Helpdesk Answer. Chr. Michelsen Institute. https://www.u4.no/publications/the-role-of-civil-society-organisations-in-countering-transnational-corruption

Massarella, K., Sallu, S. M., Ensor, J. E., & Marchant, R. (2018). REDD+, hype, hope, and disappointment: The dynamics of expectations in conservation and development pilot projects. *World Development*, 109, 375–385.

Namibian Association of CBNRM Support Organisations (NACSO). (2024). *The State of Community Conservation in Namibia: Annual Report 2023*. Windhoek: NACSO. https://www.nacso.org.na/sites/default/files/SOCC%202023%20Annual%20Report%20_%20Web%20_Repro%202.pdf

The Nature Conservancy. (2024). Partnering with Indigenous People and local communities.https://www.nature.org/en-us/about-us/who-we-are/how-we-work/community-led-conservation/

Nepali Times. (2019, January 4). Who now owns Nepal's forests? *Nepali Times*. https://nepalitimes.com/banner/who-now-owns-nepals-forests

Nepstad, Lucy & Gerber, James & Hill, Jason & Dias, Lívia & Costa, Marcos & West, Paul. (2019). Pathways for recent Cerrado soybean expansion: extending the soy moratorium and implementing integrated crop livestock systems with soybeans. *Environmental Research Letters*. 14. 10.1088/1748-9326/aafb85.

Nordhaus, W. D. (2017). Revisiting the social cost of carbon. *Proceedings of the National Academy of Sciences*, 114(7), 1518–1523. https://doi.org/10.1073/pnas.1609244114

Nordhaus, W. (2024, March 21). The Economic Consequences of Climate Change: An Interview with Nobel Laureate William Nordhaus. Yale Department of Economics. https://economics.yale.edu/news/240321/economic-consequences-climate-change-interview-nobel-laureate-william-nordhaus
Nordhaus, W. (2024, March 21). The Economic Consequences of Climate Change: An Interview with Nobel Laureate William Nordhaus. Yale Department of Economics. https://economics.yale.edu/news/240321/economic-consequences-climate-change-interview-nobel-laureate-william-nordhaus
Notre Terre Sans Pétrole Coalition. (2025, May 5). Statement condemning DRC's opening of 52 new oil blocks in the Cuvette Centrale: Oil versus forests, a destructive choice by the government. Coalition declaration and campaign materials. Available via REDD-Monitor and Rainforest Rescue: https://notreterresanspetrole.com/
Nuhu, S. (2021). Rethinking land access governance in Global South Cities : understanding the dynamics and contentions of land access processes and governance mechanisms in peri-urban areas of Dar es Salaam, Tanzania. Diss. (sammanfattning/summary) Sveriges lantbruksuniv., Acta Universitatis Agriculturae Sueciae, 1652-6880
O'Brien, J. (2025, June 11). Navigating the Waters of Peace: Challenges and Opportunities in the Implementation of Colombia's Peace Agreement. Kroc Institute for International Peace Studies, University of Notre Dame. Retrieved from ttps://kroc.nd.edu/news-events/news/navigating-the-waters-of-peace-progress-challenges-and-opportunities-in-the-implementation-of-colombias-historic-peace-agreement/
Oseland, S. E. (2019). Breaking silos: Can cities break down institutional barriers in climate planning? *Journal of Environmental Policy & Planning*, 21(4), 345–357. https://www.tandfonline.com/doi/abs/10.1080/1523908X.2019.1623657
Ostrom, E. (1990). *Governing the Commons: The Evolution of Institutions for Collective Action*. Cambridge University Press.
Ostrom, E. (2009). *Understanding Institutional Diversity*. Princeton University Press.
Pagiola, S., Honey-Rosés, J., & Freire-González, J. (2016). Evaluation of the permanence of land use change induced by payments for environmental services in Quindío, Colombia. *PLOS ONE*, 11(3), e0147829. https://doi.org/10.1371/journal.pone.0147829
Pascual, U., Adams, W. M., Díaz, S., Lele, S., Mace, G. M., & Turnhout, E. (2021). Biodiversity and the challenge of pluralism. *Nature Sustainability*, 4, 567–572. https://doi.org/10.1038/s41893-021-00694-7
Pielke, R. A. (2007). *The Honest Broker: Making Sense of Science in Policy and Politics*. Cambridge University Press.
Plurinational State of Bolivia. (2012). Ley N° 300 Marco de la Madre Tierra y Desarrollo Integral para Vivir Bien [Framework Law of Mother Earth and Integral Development for Living Well]. Enacted October 15, 2012. Retrieved from https://www.planificacion.gob.bo/uploads/marco-legal/Ley%20N%C2%B0%20300%20MARCO%20DE%20LA%20MADRE%20TIERRA.pdf
Prabowo, H. Y., & Cooper, K. (2017). Re-understanding corruption in the Indonesian public sector through three behavioral lenses. *Journal of Financial Crime,* 23(4), 1028–1062. https://www.sciencedirect.com/org/science/article/abs/pii/S135907901600035X
Proforest & Landesa. (2023). Respecting rights of Indigenous Peoples and local communities in landscape initiatives: A guide for practitioners on minimum safeguards and evolving best practices (Version 1.0, March 2023). *Proforest*. https://www.proforest.net/fileadmin/uploads/proforest/IPLCs_in_Landscape_Initiatives.pdf
Pyles, M. V., Magnago, L. F. S., Maia, V. A., Pinho, B. X., Pitta, G., de Gasper, A. L., Vibrans, A. C., dos Santos, R. M., van den Berg, E., & Lima, R. A. F. (2022). Human impacts as the main driver of tropical forest carbon. *Science Advances*, 8(6), eabl7968. https://doi.org/10.1126/sciadv.abl7968
Rainforest Alliance. (2023). Maximizing impact in critical landscapes. In 2023 Annual Report. Retrieved from https://www.rainforest-alliance.org/annual_report/2023/

Rainforest Foundation Norway. (2021). Indigenous people receive little climate funding. Oslo: Rainforest Foundation Norway. Retrieved from https://www.regnskog.no/en/news/falling-short

Rainforest Foundation UK. (6 May 2025). DRC government opens 52 oil blocks in the Cuvette Centrale peatlands, undermining green commitments. https://www.rainforestfoundationuk.org/drc-government-opens-52-oil-blocks-in-the-cuvette-centrale-peatlands-undermining-green-commitments/

Rainforest Foundation UK. (2025a, July 29). Key Land-Use Planning Law Passed in the DRC. https://www.rainforestfoundationuk.org/key-land-use-planning-law-passed-in-the-drc/

Rainforest Foundation US. (2024). 2024 Annual Report: Progress at Rainforest Foundation US. Retrieved from https://rainforestfoundation.org/about/financials-transparency/2024-annual-report/

Rainforest Rescue. (2025). DRC approves oil exploration in critical peatland. https://www.rainforest-rescue.org/updates/13654/drc-approves-oil-exploration-in-critical-peatland

Reed, J., Ickowitz, A., Chervier, C., Djoudi, H., Moombe, K., Ros-Tonen, M., Yanou, M., Yuliani, L., & Sunderland, T. (2020). Integrated landscape approaches in the tropics: A brief stock-take. *Land Use Policy*, 99, 104822. https://www.sciencedirect.com/science/article/pii/S0264837719322999

Remme, M., McGuire, F., Guthrie, T., Yagnik, P., Small, R., Webb, D., & Arquitt, S. (2019). Financing across sectors for sustainable development: Guidance note. *United Nations Development Programme (UNDP)*. https://www.undp.org/sites/g/files/zskgke326/files/publications/Guidance%20Note_Financing%20across%20sectors%20for%20sustainable%20development.pdf

Reytar, K., Veit, P., & von Braun, J. (2024). Protecting Biodiversity Hinges on Securing Indigenous and Community Land Rights. World Resources Institute. https://www.wri.org/insights/indigenous-and-local-community-land-rights-protect-biodiversity

Ribot, J. C., & Larson, A. M. (Eds.). (2012). *Democratic Decentralization through a Natural Resource Lens: Cases from Africa, Asia and Latin America*. Routledge/Earthscan.

Ribot, J. C., & Peluso, N. L. (2003). A theory of access. *Rural Sociology*, 68(2), 153–181. https://onlinelibrary.wiley.com/journal/15490831

Rights and Resources Initiative. (2015). Who owns the world's land? A global baseline of formally recognized Indigenous and community lands.

Rights and Resources Initiative. (2018). *A Global Baseline of Carbon Storage in Collective Lands*. Washington, DC: Rights and Resources Initiative.

Rights and Resources Initiative. (2020a). The Opportunity Framework 2020: Identifying Opportunities to Invest in Securing Collective Tenure Rights in the Forest Areas of Low- and Middle-Income Countries. Technical Report. https://rightsandresources.org/wp-content/uploads/Opp-Framework-v2021.pdf

Rights and Resources Initiative. (2023). At a crossroads: Trends in recognition of community-based forest tenure from 2002–2023. https://rightsandresources.org/publication/at-a-crossroads-2023/

Rights and Resources Initiative & Rainforest Foundation Norway. (2024, June 26). New Online Dashboard Brings Transparency to Climate Finance; Gives Open Access to Funding Data for Indigenous Peoples', Local Communities', and Afro-descendant Peoples' Tenure Rights and Forest Guardianship. Rights and Resources Initiative Blog. https://rightsandresources.org/blog/new-online-path-to-scale-dashboard/

Rijk, G., Warmerdam, W., & Kuepper, B. (2023, July). €0.7 billion in profits, €66 billion in damages: Rabobank's destructive financing of deforestation in Brazil. *Profundo for Greenpeace Netherlands*. Retrieved from: https://www.greenpeace.org/static/planet4-netherlands-stateless/2023/07/e70522eb-profundo_rabobank-brazil-final-july2023.pdf

Rocky Mountain Institute (RMI). (2023, December 19). *From Paper to People: Bringing Equity to Carbon Markets*. RMI. https://rmi.org/from-paper-to-people-bringing-equity-to-carbon-markets/

Sahide, M. A. K., & Giessen, L. (2015). The fragmented land use administration in Indonesia – Analysing bureaucratic responsibilities influencing tropical rainforest transformation systems. *Land Use Policy*, 49, 504–512. https://doi.org/10.1016/j.landusepol.2015.08.018

Sarmiento-Barletti, J. P., Larson, A. M., van Oosten, C., & Guerrieri, L. (2022). Understanding systemic land use dynamics in conflict-affected territories in Colombia. *PLOS ONE*, 17(6), e0269088. https://doi.org/10.1371/journal.pone.0269088

Sawathvong, S., & Hyakumura, K. (2024). A comparison of the Free, Prior, and Informed Consent (FPIC) guidelines and the "implementation of governance, forest landscapes, and livelihoods" project in Lao PDR: The FPIC team composition and the implementation process. *Land*, 13(4), 408. https://www.mdpi.com/2073-445X/13/4/408

SCALA/UNDP. (2025). *Programme Highlights Report 2024*. United Nations Development Programme and Food and Agriculture Organization. https://www.adaptation-undp.org/resources/scala-programme-highlights-report-2024

Schlehe, J. (2024, January 31). Coal vs Climate – Indonesia's Energy Policy Contradicts Its Climate Goals. GIGA Focus Asia. https://www.giga-hamburg.de/en/publications/giga-focus/coal-vs-climate-indonesia-s-energy-policy-contradicts-its-climate-goals

SOMO & Kenya Human Rights Commission (KHRC). (2023). *The Dark Side of Green Carbon: The Kasigau Corridor REDD+ Project and Human Rights Challenges in Kenya*. Centre for Research on Multinational Corporations (SOMO) and Kenya Human Rights Commission. Retrieved from https://www.somo.nl/the-dark-side-of-green-carbon-the-kasigau-corridor-redd-project-and-human-rights-challenges-in-kenya/

Stechemesser, A., Koch, N., Mark, E., Dilger, E., Klösel, P., Menicacci, L., Nachtigall, D., Pretis, F., Ritter, N., Schwarz, M., Vossen, H., & Wenzel, A. (2024). Climate policies that achieved major emission reductions: Global evidence from two decades. *Science*, 385(6711), 884. https://doi.org/10.1126/science.adl6547

Sunderlin, W. D., de Sassi, C., Sills, E. O., Duchelle, A. E., Larson, A. M., Resosudarmo, I. A. P., & Sun, Y. (2018). Creating an appropriate tenure foundation for REDD+: The record to date and prospects for the future. *World Development, 106*, 376–392. https://doi.org/10.1016/j.worlddev.2018.01.010

Sustainability Directory. (2025). *Sustainability Directory: ESG, Supply Chains, and Sustainability Insights*. Retrieved July 25, 2025, from https://sustainability-directory.com/

Tauli-Corpuz, V., Alcorn, J., Molnar, A., Healy, C., & Barrow, E. (2020). Cornered by protected areas: Adopting rights-based approaches to enable cost-effective conservation and climate action. *World Development*, 130, 104923. https://www.sciencedirect.com/science/article/pii/S0305750X20300498

Tenure Facility. (2022). Scaling up the community land value chain (CaVaTeCo) approach in Mozambique. https://thetenurefacility.org/projects/scaling-up-the-community-land-value-chain-cavateco-approach-in-mozambique/

The Nature Conservancy (2022). The Voice, Choice, and Action Framework: A Conservation Practitioner's Guide to Indigenous and Community-Led Conservation, Version 2.0. Arlington, VA.

Transparency International. (2024). Corruption perceptions index 2023. https://www.transparency.org/en/cpi/2023

Turnhout, E., Dewulf, A., & Hulme, M. (2016). What does policy-relevant global environmental knowledge do? The cases of climate and biodiversity. *Current Opinion in Environmental Sustainability*, 18, 65–72. https://doi.org/10.1016/j.cosust.2015.09.004

United Nations. (2007). United Nations Declaration on the Rights of Indigenous Peoples (UNDRIP). A/RES/61/295. Adopted by the UN General Assembly on 13 September 2007. https://www.un.org/development/desa/indigenouspeoples/wp-content/uploads/sites/19/2018/11/UNDRIP_E_web.pdf

United Nations Department of Economic and Social Affairs (UN DESA). (2024). *HIGHLIGHTS 2023–2024: Towards Sustainable Development for All*. New York: United Nations. https://www.un.org/sites/un2.un.org/files/desa-highlights-report-2023-2024.pdf

United Nations Development Programme (UNDP) Ghana. (2024, September 5). *Empowering Communities for Sustainable Resource Management*. UNDP. https://www.undp.org/ghana/press-releases/empowering-communities-sustainable-resource-management

U.S. Agency for International Development (USAID). (2022). Localize the Approach: USAID's New Commitment to Localization. https://www.usaid.gov/localization

van Bommel, N., & Höffken, J. I. (2023). The urgency of climate action and the aim for justice in energy transitions – dynamics and complexity. *Environmental Innovation and Societal Transitions*, 48, 100763. https://doi.org/10.1016/j.eist.2023.100763

van Noordwijk, M., Tomich, T., & Verbist, B. (2001). Negotiation support models for integrated natural resource management in tropical forest margins. *Conservation Ecology*, 5(2), 21. http://www.consecol.org/vol5/iss2/art21/

Villavicencio-Calzadilla, P. (2025, January 29). The Rights of Mother Earth in Bolivia: Progress and Challenges. Case Study. Heinrich Böll Stiftung. https://www.boell.de/en/2025/01/29/rights-mother-earth-bolivia-progress-and-challenges

Voosen, P. (2025, October 11). Trump seeks to end climate research at premier U.S. climate agency. *Science* | AAAS. Retrieved from https://www.science.org/content/article/trump-seeks-end-climate-research-premier-u-s-climate-agency

Weiner, J. F., & Glaskin, K. (2007). Customary Land Tenure and Registration in Papua new Guinea and Australia: Anthropological Perspectives. In J. F. W. A. K. G. (Ed.), *Customary Land Tenure and Registration in Australia and Papua New Guinea: Anthropological Perspectives* (1st ed., Vol. 1, pp. 1–14). ANU ePress.

Wily, L. A. (2018). Collective Land Ownership in the 21st Century: Overview of Global Trends. *Rights and Resources Initiative*. Retrieved from https://rightsandresources.org/wp-content/uploads/2018/06/Collective-Land-Ownership-in-the-21st-Century_Liz-Alden-Wily_May-2018.pdf

WOLA (Washington Office on Latin America). (2019, August 27). *Fact Sheet: The CICIG's Legacy in Fighting Corruption in Guatemala*. WOLA. https://www.wola.org/analysis/cicigs-legacy-fighting-corruption-guatemala/

World Bank. (2024). Improved forest and landscape management in the Democratic Republic of the Congo (DRC). https://www.worldbank.org/en/news/feature/2024/07/31/improved-forest-and-landscape-management-afe-democratic-republic-of-the-congo-drc

World Bank Development Finance Vice Presidency. (2022, July 19). *Understanding Trends in Proliferation and Fragmentation for Aid Effectiveness During Crises*. Washington, DC: World Bank Group. https://thedocs.worldbank.org/en/doc/ef73fb3d1d33e3bf0e2c23bdf49b4907-0060012022/understanding-trends-in-proliferation-and-fragmentation-for-aid-effectiveness-during-crises

World Resources Institute (WRI). (2023, January 6). Indigenous Peoples and local communities: Helping people protect their lands and the many benefits they provide. https://www.wri.org/equitable-development/indigenous-peoples-local-communities

World Wide Fund for Nature (WWF). (2016). Biodiversity, People and Climate Change: Final technical report of the Hariyo Ban Program, first phase.

WWF Namibia. (2023). *WWF Namibia Impact Report 2023*. Windhoek: WWF Namibia. https://wwfafrica.awsassets.panda.org/downloads/2023-wwf-impact-report-21-dec-23-compressed.pdf

7 Shifting From Crisis to Credibility

Negotiated Governance in Climate Mitigation

> Only a crisis – actual or perceived – produces real change. When that crisis occurs, the actions that are taken depend on the ideas that are lying around. That, I believe, is our basic function: to develop alternatives to existing policies, to keep them alive and available until the politically impossible becomes politically inevitable.
>
> – Milton Friedman, Preface to *Capitalism and Freedom*, 1982

> We need solutions that match the scale of the challenge and a new politics to get them done. Both are well overdue.
>
> – Tony Blair, Foreword to *The Climate Paradox: Why We Need to Reset Action on Climate Change*, Fursman (2025)

Whatever one thinks of Friedman's broader ideology and economic theories, this observation remains strikingly relevant today. The perception of crisis, real or imagined, goes far beyond climate issues in 2025.

In the land sector of the Global South, climate mitigation is facing a credibility crisis. That crisis is not merely ecological. It is institutional, political, conceptual, and strategic. It is a failure to translate global climate ambition into evidence-supported, appropriate, and feasible strategies that also build on biodiversity, livelihoods, and governance realities (3ie, 2024) of the landscapes where mitigation is meant to occur. It is a failure to place the right tools in the right people's hands, constraining hope of achieving sought-after 'triple-wins' (CCAFS, 2017), as IPLCs remain largely sidelined in high level planning (Cultural Survival, 2022).

Despite decades of reform rhetoric, the fundamental structures of decision-making remain centralized, top-down, and externally driven. What continues to pass as innovation – new carbon accounting methodologies, high-integrity offsets, participatory safeguards – is too often a rebranding of the same underlying model: technocratic design, counterfactual baselines, externally imposed metrics, and superficial forms of participation. Old wine in new bottles.

At the same time, thought leaders like Tony Blair and others at the Blair Institute for Global Change (Fursman, 2025) position the global community as moving from the Era of Apathy we currently are in into the Era of Disruption, where Nature-based Solutions (NbS) must assume a critical transitional or bridging role while decarbonization and engineered CDR technologies are deployed and scaled

DOI: 10.4324/9781003380320-8

(Romm et al., 2025). Specifically, "some of the best near-term returns on investment will be through nature-based solutions that will need support to accelerate." As discussed throughout this book, NbS may have a solid ecological evidence base and has great intuitive appeal, but the institutional and social feasibility side remains critically underdeveloped. Ambiguous net-zero targets under the Paris Agreement (Rogelj et al., 2021) do not facilitate in moving forward.

Persistent deforestation, limited climate finance reaching the ground, and underwhelming progress on governance and the Sustainable Development Goals (SDGs) (FAO and UNEP, 2020; UNDP, 2014; UNEP, 2022) reflect the same deeper pattern: a mismatch between internationally imposed frameworks and local contexts. Despite two decades of experimentation – first with REDD+, now increasingly with NbS – the dominant paradigm continues to rely on top-down models, technocratic assumptions, and consultation processes that too often bypass or marginalize the very communities whose participation is essential. In this impending period of disruption, the global community, driven by international institutions with the greatest leverage, is poised to rely on conventional wisdoms that must in themselves be rethought as part of a new paradigm for how NbS, Natural Climate Solutions, and their components can play a positive disruptive role in CMLSGS.

This book has argued throughout that the real problem is not that REDD+ or NbS are inherently flawed in aspiration. The issue is how they have been implemented through a prevailing set of strategies, tools, approaches, and methods (STAM) that remain misaligned with the social, institutional, and ecological conditions of the Global South. When these STAM frameworks fail to recognize who governs land, how decisions get made, or what incentives are meaningful, they not only underperform – they often cause harm, alienating local actors, undermining tenure security, and producing short-lived or extractive outcomes. The failure to apply a substantivist economic anthropology lens to the challenge, as discussed in earlier chapters, perpetuates implausible "solutions."

This final chapter does not repeat the diagnosis. Instead, it takes seriously the opportunity embedded in the crisis. Suppose we accept that credibility and sustainability cannot be delivered through external design alone. In that case, the imperative is clear: We must begin from the ground up, with a negotiated framework that elevates IPLCs as leaders in integrated land use management decision making, that centers IPLCs as leaders and drivers of solutions as the UNFCCC LCIPP stipulates (2024), not just stakeholders or beneficiaries. Any immediate and systemic mitigation and adaptation called for by the IPCC (2022) will only work if IPLCs become fully empowered.

This shift cannot be based on slogans or virtue signaling. It must begin with tenure security and move through capacity building toward structured negotiation. In many parts of the Global South IPLCs possess deep traditional knowledge and have shown strong advocacy capacity but often lack access to the information, analytical tools, or institutional support required to enter high-stakes negotiations on equal footing. Advocacy capability is not the same as negotiation readiness. That gap needs to be acknowledged – not to dismiss their agency, but to enable it.

For negotiated governance to work, social license must first be earned, not presumed. Information must be accessible, and costs, risks, and trade-offs must be assessed transparently by those who own and use land and forest - not external agencies on their behalf. This is not a call for indefinite delays, nor does it excuse government inaction. Rather, it emphasizes that capacity building is not a preliminary stage that precedes negotiation – it must be part of any negotiation process itself. When structured well, negotiation builds both legitimacy and ownership. It allows IPLCs, governments, and the private sector to jointly define the terms of engagement and clarify what "success" actually means in a given place.

None of this implies that IPLCs alone can solve the climate crisis. They cannot and should not be burdened with that expectation. National governments remain essential – both in providing legal authority and in managing the regulatory and enforcement frameworks without which even the most promising community-led efforts will falter. What this chapter advocates is a rebalancing: a recognition that IPLC leadership is a matter not only of justice but also of practical necessity.

The current moment – with REDD+ under scrutiny, and NbS and NCS poised to repeat many of the same mistakes – presents an opening. It is time to move past the illusion that projectized consultation and externally controlled finance will produce scalable or sustainable outcomes. What is needed is not a new acronym or policy directive. What is needed is a political and institutional reset.

From Principles to Practice: Empowering IPLCs Through Integrated, Negotiated Governance

While the principle of putting people first is now echoed across donor rhetoric and high-level communiqués, it remains elusive in practice. This chapter argues that a genuine shift toward integrated and sustainable climate mitigation in the Global South must start by operationalizing IPLCs as co-governors, not just as stakeholders. But achieving this requires far more than symbolic gestures. It demands tenure security, investment in capacity, and, above all, negotiated platforms that recognize IPLCs as legitimate actors with the authority to shape land-use and climate decisions.

Chapter 6 introduced the integrative strategy of Negotiated Governance Platforms (NGPs). NGPs must be anchored in "Good Enough Land Tenure Security" (GELTS) as a non-negotiable precondition, a principle discussed in Chapter 4. As demonstrated in Ethiopia's rural land certification program that secured 20 million certificates issued via decentralized committees and Kenya's Community Land Act reforms that succeeded in formalizing collective tenure for pastoralist communities, both discussed in Chapter 4, GELTS can provide the minimal tenure clarity required for IPLCs to engage in negotiations without fear of dispossession. Without this foundation, NGPs risk replicating the exclusionary dynamics seen in Cambodia's Economic Land Concessions, where weak tenure and corruption enabled elite capture through fraudulent permits and coerced displacement (see Chapter 4: Political and Economic Barriers to Land Tenure Reform; Cambodia Timber Sector Case Study), or in the Brazilian cerrado where land grabbing continues (Nepstad et al., 2019; Global Witness, 2024).

The rationale is evidence-based. IPLCs manage or claim customary tenure over large swaths of the world's remaining intact forests – areas essential to both climate stability and biodiversity (Garnett et al., 2018; Rights and Resources Initiative,

2023). Their stewardship, where protected by secure land rights, has consistently resulted in lower deforestation rates and greater ecological resilience than comparable lands under state or private control (Boucher et al., 2014; World Resources Institute, 2024). As Dawson et al. (2024) emphasize, equitable governance based on shared or primary control by IPLCs correlates with more sustainable ecological outcomes. Yet despite this track record, IPLCs continue to be excluded from the design and governance of most climate mitigation frameworks.

The default process endorsed by international donors and multilateral actors has been consultation, presented as a best practice and often as a proxy for FPIC. The IPBES and IPCC, however, recognize the pitfalls of top-down consultation in national implementation and global agreements, and have called for inclusive, collaborative, and multi-level approaches engaging communities, Indigenous Peoples, and other local actors (Pörtner et al., 2021). Even when digital tools are deployed, they rarely foster the kind of deliberative engagement required for communities to weigh trade-offs and negotiate terms. Worse, as shown in numerous REDD+ projects, "consultation" often functions as procedural camouflage for elite capture (Angelsen et al., 2018; Carbon Market Watch, 2023).

While many IPLCs have strong advocacy experience, community cohesion, and deep place-based knowledge, this does not automatically translate into the analytical and technical capabilities required to engage in complex climate finance or land-use planning negotiations. In sub-Saharan Africa especially, the asymmetries in legal literacy, data access, and institutional backing are stark – and must be addressed before meaningful negotiation can begin.

This aligns with newer calls for hybrid governance that blends traditional knowledge with state and scientific systems. Berkes et al. (2000) decades ago introduced adaptive co-management and blended knowledge systems to drive conservation policy and practice. Building on this, we argue that negotiation platforms should incorporate participatory cost-benefit frameworks, Geographic Information Systems-enabled land-use planning, and conflict resolution protocols grounded in customary norms. New technologies, including AI-driven land suitability tools and spatial analytics, can further enhance community capacity to navigate increasingly complex planning environments – *if* designed inclusively and with community consent.

Crucially, negotiation cannot proceed in the absence of social license. While the term is often used informally, it refers here to the legitimacy and consent required to implement policies or investments at the landscape level. In many cases, communities are unaware of climate mitigation plans that directly impact them, including offset schemes linked to Article 6 or jurisdictional carbon deals negotiated at the national level. Even when consultation occurs, it may reflect elite consensus or superficial participation. Without transparency, trust, and deliberate effort to generate legitimacy from below, negotiated governance becomes just another imposed model.

This problem is exacerbated by opacity at the national level. Governments may enter into agreements with donors or corporations that are shielded from public view, heightening perceptions of corruption and eroding local trust. These dynamics undermine not only the feasibility of climate action but also its long-term credibility. As numerous examples have shown – from Kenya's Kasigau REDD+

project (SOMO & KHRC, 2023) to the Democratic Republic of Congo's community forestry programs (ClientEarth, 2023) – lack of transparency opens the door to both elite capture and contested land claims.

If climate mitigation is to succeed in the Global South, negotiation must replace consultation, and that negotiation must be capacitated, accountable, and rooted in legitimacy. This is not a claim that IPLCs are always fully ready to assume leadership in every context. Rather, it is a call to build toward that readiness deliberately. Structured investments in land tenure security, analytical training, legal support, and platform governance are not add-ons – they are prerequisites.

From Simplistic "Triple Wins" to Transformational Pathways

For decades, the notion of achieving "triple wins" (CCAFS, 2017; Suckall et al., 2015) – simultaneous gains in poverty alleviation, agricultural development, and climate change mitigation – has shaped much of the Global North's climate and development programming in the Global South. This logic underpins initiatives like Climate-Smart Agriculture (CSA), which aims to enhance productivity, build resilience, and reduce greenhouse gas emissions (FAO, 2023; World Bank, 2024). On paper, these strategies appear elegant: Who could argue with improving productivity, protecting ecosystems, and reducing emissions at the same time?

Yet the practical realities are more complex. Reviews have shown that triple-win projects often face significant implementation challenges due to underlying power asymmetries and conflicting interests (CCAFS, 2017; Future Agricultures, 2013). Elite capture, insecure tenure, and poorly designed participation processes frequently undermine the promise of equitable, sustainable outcomes. While the rhetoric of synergy persists, the actual distribution of benefits often favors external stakeholders, while local communities absorb the costs.

This pattern is especially evident in the unequal flows of climate finance and the commodification of carbon. While triple-win rhetoric often includes a call for "just transitions," the historical climate debt owed by the Global North to the Global South continues to go unaddressed in structural terms (Climate Network, 2024). The implementation of mitigation projects in the name of triple wins often reproduces old hierarchies under new labels, reinforcing the very systems that climate justice seeks to challenge.

That said, there are cases – particularly in decentralized renewable energy and regenerative agroecology – where overlap between mitigation, adaptation, and development has proven more than theoretical. As Project Drawdown (2023) emphasizes, what distinguishes these initiatives is not the elegance of the concept but the mode of implementation. When projects are community-led and science-informed, they are far more likely to yield integrated, enduring results.

This book argues that the path forward is not to abandon the pursuit of multiple benefits, but to transform the way those benefits are identified, negotiated, and shared. Real "triple wins" that are nature-positive and premised on growth (WEF, 2024) are not declared in policy memos – they are built through process. That means putting IPLCs at the negotiating table, not merely listing them as beneficiaries.

This requires a shift in framing – from NbS and NCS carbon-centered models of "success" to one based on negotiated governance among IPLCs, governments, and private sector actors. Drawing on Bebbington and Humphreys Bebbington's (2011) analysis of the political and institutional complexities shaping land use in countries experiencing the "resource curse," and echoing Sunderlin et al. (2014), sustainable land governance must explicitly engage with overlapping legal and customary rights, diverse economic interests, and the long-term incentives of all parties. None of that can be done through top-down program templates. Nor does current NbS and NCS "best practice" offer solid evidence-based direction.

The Role of International Organizations: Dual Forces of Support and Distortion

International organizations (IOs) have played contradictory roles in this evolution. On the one hand, many have championed communities' rights and supported tenure security, biodiversity co-management, and procedural safeguards like FPIC. On the other hand, their structural incentives often reinforce centralized control, efficiency-driven timelines, and standardized implementation. These tendencies constrain the space for IPLC-led negotiation and long-term institutional strengthening.

IOs also tend to promote their own toolkits and preferred methodologies, often crowding out grassroots solutions or national processes that take longer to develop but may be more legitimate and contextually appropriate. As a result, even when they speak the language of empowerment, their operational frameworks tend to favor large implementing NGOs or private firms with little embeddedness in local contexts.

Transforming this dynamic will require IOs to evolve from technical enforcers to institutional enablers. That means creating space – financially, politically, and procedurally – for tripartite negotiation platforms where IPLCs, government actors, and external stakeholders can collectively define land-use strategies that are not only technically sound but also politically and socially legitimate. Only then can the seductive logic of "triple wins" become something more than an aspirational slogan. To state the obvious: Getting there won't be easy.

From Retrenchment to Rebalancing: Why the Shift Is No Longer Optional

For nearly two decades it has been fashionable for many IOs, from the Global North, increasingly discuss transitioning from program drivers to support agencies. As the Center for Global Development wryly observes, the development sector has long aspired to "work itself out of a job" – a sentiment frequently voiced, yet rarely enacted (Center for Global Development, 2017). Given the mounting evidence that the current paradigm isn't delivering – either for people or for the planet – this evolution is overdue.

The Intergovernmental Science-Policy Platform on Biodiversity and Ecosystem Services (IPBES) found that one million species are at risk of extinction and that 75% of terrestrial and two-thirds of marine ecosystems have already been significantly altered by human activity (IPBES, 2019). At the same time, the UN's 2023

SDGs Report warns that over 600 million people – especially in rural areas of the Global South – are projected to remain in extreme poverty by 2030, with multiple development indicators reversing course in recent years (United Nations, 2023). In this context, empowering IPLCs through NGPs is not just desirable. It is essential.

And as for "working ourselves out of a job," one might say the US Government's Department of Government Efficiency took that slogan a bit too literally – though rather than gradually stepping back to empower local systems, it seems they opted to repurpose it into a line item for dismantling USAID wholesale as an institution, along with over 85% of its budget. This was not because local systems were finally ready to lead, but because global development was politically rebranded as expendable (Galvin, 2025). As Elon Musk put it, this was "a time to die" for US development assistance, not because the mission was accomplished, but because the politics of multilateralism had collapsed (Jalali, 2025).

For land-sector climate mitigation in the Global South, this only sharpens the case for structural realignment as part of a broader paradigm shift. As the financial and institutional underpinnings of traditional donor-led models weaken, negotiated, locally grounded governance must become the norm – not just an ideal, but also an urgent necessity for local stewards and their governments.

Nor Are Safeguards the Solution

Despite decades of rhetoric and good intentions concerning safeguards like FPIC and community consultation, these mechanisms have largely failed to transfer meaningful decision-making power to IPLCs. While necessary, the recourse to cite safeguards (along with co-benefits) as central to the logic of climate mitigation projects (along with conservation projects, which are part of the NbS arsenal) has proven hollow with many IPLCs, as noted throughout this book. Safeguards may prevent the worst harm, but they do little to promote IPLCs' legitimate sustainable development aspirations.

Too often, consultation processes remain tokenistic, poorly sequenced, or manipulated to legitimize predetermined outcomes, even if safeguards are involved. A case study from the Philippines illustrates this dynamic starkly: FPIC was described as "a state mechanism to authorize or ratify the displacement of Indigenous peoples . . . and legitimize the plunder of their wealth by big business" (Daytec-Yañgot, 2021). In the case of REDD+ this has led to all too frequent IPLC rights abuses (Sarmiento Barletti and Larson, 2017).

Dudine and Szoke-Burke (2020) argue that FPIC has too often been reduced to a perfunctory checkbox. Loose interpretations of "informed" and "consent" continue to undermine its intent – an issue I raised in *Redeeming REDD* over a decade ago (Brown, 2013), and one that would merit doctoral dissertations.

There are, of course, efforts pointing in a more promising direction. The Dedicated Grant Mechanism for Indigenous Peoples and Local Communities (DGM) – supported by the Climate Investment Funds – has delivered over $80 million to IPLC-led programs. Yet this remains a fraction of the global landscape. Less than 1% of climate finance directly reaches such initiatives (CIF, 2023).

Even well-intentioned programs by BINGOs and multilateral institutions continue to position IPLCs as beneficiaries, not as architects or negotiators of solutions. This marginalization is especially concerning given the scale of community-managed territories. Research by the Rights and Resources Initiative (2020) estimates that at least 32% – and possibly up to 50% – of the world's land is held or managed by IPLCs. The World Bank, meanwhile, acknowledges that Indigenous Peoples manage roughly a quarter of the world's land, including a large share of protected areas. Whatever the number, IPLC leverage over the future of the land sector of the Global South is massive.

The Forest Declaration Assessment goes further, suggesting IPLCs provide ecosystem services worth more than US$ 1.16 trillion annually. These lands are often among the most biodiverse and carbon-rich on Earth (Schleicher et al., 2017). Yet in most countries, customary lands lack formal legal recognition. Tenure insecurity remains a major obstacle to climate-compatible, socially feasible mitigation.

Yet the institutional logic of climate governance has rarely allowed for shifting to socially feasible mitigation. While many studies highlight the conservation benefits of Indigenous governance (e.g., Artelle et al., 2019), few have grappled with what it would mean for IPLCs to negotiate directly with governments and private actors. Policy discourse still tends to focus on rights in the abstract, rather than on the design of functional power-sharing mechanisms (Agrawal & Chhatre, 2006). Moreover, the emphasis on safeguards and technocratic FPIC creates an illusion of participation and ownership. This book attempts to fill that gap by outlining how negotiated leadership can – and should – be structured.

Preceding chapters argued for redefining success in CMLSGS through practical metrics: appropriateness, feasibility, tangible outcomes, and long-term sustainability. These must become the North Star guiding climate investments and programming. When this redefinition is embraced, the private sector can also be leveraged in more accountable and constructive ways.

TFFF as Contributor to a Potential Negotiated Governance Pathway Forward?

The Tropical Forest Forever Facility (TFFF), launched by Brazil at COP28 and aiming for full operationalization by COP30 in Belém, represents a rare attempt at reimagining how tropical forest conservation might be financed at scale. Touted as "a forest finance tool like no other" (Streck, 2024), TFFF aspires to provide predictable, long-term funding through direct payments to tropical rainforest countries, based not on abstract carbon accounting but on actual hectares conserved or restored. As of late 2025, many TFFF critics disagree (Global Forest Coalition & Fundación Solón, 2025; Global Witness, 2025; World Rainforest Movement, 2025).

With a fundraising target of $125 billion, the fund aims to deliver predictable, long-term financial resources, emphasizing simplicity, performance-based incentives, and measurable area-based outcomes rather than carbon market offsets. This design, if grounded in real transparency and community authority, offers a potential departure from REDD+ and similar NbS platforms that have been criticized for weak safeguards and unverifiable claims (Carbon Market Watch, 2023).

TFFF explicitly includes a commitment to IPLC participation – at least 20% representation is promised – but this floor should be treated as a starting point, not a ceiling. For TFFF to reflect the paradigm shift advocated in this book, three structural changes are essential:

1. IPLC Co-Design and Co-Governance: IPLCs must be centrally involved in designing how TFFF operates at the national and subnational levels. This means not only FPIC in principle but also negotiated decision-making roles in setting performance metrics, designing verification systems, and determining eligibility criteria.
2. Simplified Metrics and Measurable Outcomes: TFFF's move away from complex carbon metrics is positive. Measuring forest area conserved is more operationally feasible and reduces the barriers IPLCs face in participating meaningfully. But these metrics must be co-developed with communities to reflect social as well as ecological success.
3. Direct Financing and Long-Term Incentives: Like many climate finance mechanisms, TFFF's success hinges on bypassing intermediaries and ensuring that funding flows directly to those managing the land.

To avoid repeating REDD+'s cycle of aspirational rhetoric and underwhelming delivery – where IPLCs are framed as key beneficiaries but are often sidelined in practice – TFFF must commit to genuine co-design with them as it purports to (Henneberger, 2025). This is the only way to achieve scale in remote landscapes and become a model for the next generation of more successful conservation and climate mitigation programming. This includes ensuring that predictable, long-term payments are linked to clearly defined land stewardship agreements, negotiated transparently with full clarity on transaction costs, timelines, and benefit-sharing terms.

TFFF could help illustrate a new path, not because it solves all problems, but because it creates space to redefine how large-scale conservation finance is structured. If linked to jurisdictional negotiation platforms that formalize IPLC–state–private sector dialogues, and if embedded in land tenure security systems, it could become part of a broader transformation toward integrated governance.

But if implemented top-down, without correcting power imbalances, addressing land tenure insecurity, or engaging IPLCs as rights-holders and architects, it will likely replicate the very problems it claims to fix, representing yet another "false solution" (Malig & Solón, 2025).

The Centrality and Limits of IPLCs in CMLSGS

As we look to alternative models for CMLSGS, it's important to acknowledge both the strategic centrality of IPLCs *and* the practical limitations that must be addressed if their role is to move beyond rhetoric. The goal isn't to romanticize IPLCs or assign them tasks they cannot reasonably be expected to perform. They are not the silver bullets. They are, however, central and continue to assume a marginalized role that simply perpetuates ineffective programming and far too frequent failure.

Indigenous Peoples are extraordinarily diverse, and local communities that do not claim or enjoy the status of indigeneity are more numerous and equally diverse. While some groups are well prepared to actively participate in all aspects of climate mitigation, many are not yet positioned to do so. Either way, they cannot, and should not, be asked to take over the functions of state enforcement, anti-corruption oversight, addressing illegality in all its forms, or international climate finance governance. Nor are they uniformly prepared, at this stage, to step into complex negotiations with donors, private actors, or government ministries operating with full legal teams and decades of experience.

Still, across the Global South, many IPLCs remain locked in a structural disadvantage. In places like sub-Saharan Africa, dual land systems persist: The state often retains de jure ownership, while IPLCs assert customary claims that are, at best, unevenly recognized or operationalized. These communities have often shown themselves to be skilled advocates, capable of mobilizing, organizing, and making their voices heard. But advocacy doesn't necessarily equate to analytical capacity, legal preparedness, or fluency in the technical and economic models underpinning today's climate deals.

This is where the paradigm needs to shift. Expecting IPLCs to negotiate equitable deals without meaningful support is not empowerment – it's entrapment. If IPLCs are going to be treated as serious players in climate mitigation, versus paternalistically, they need more than invitations to consultation meetings. They need tools: access to data, options analysis, legal support, and practical ways to evaluate risks and trade-offs. In that sense, capacity building isn't an accessory to negotiated governance – it's a precondition for it.

This is especially important given how polarized and dogmatic the policy space has become. Despite a growing evidence base on IPLC effectiveness, programming continues to reflect hardened presuppositions, particularly around the centrality of financializing nature. The prevailing logic assumes that commodifying ecosystem services is not only possible but also inevitable. Yet this view is often imposed in places where nature is not seen – or treated – as a commodity by the people who live with it. And when nature-based strategies built on this logic collide with local realities, the result isn't transformation. It's friction. Or failure.

What continues to pass as innovation in global circles – carbon credit schemes, "high-integrity" offsets, tokenistic safeguards – often boils down to the same underlying model: top-down planning, implausible counterfactual baselines, externally imposed metrics, and a persistent overestimation of how often FPIC is meaningfully achieved rather than merely claimed. What is missing isn't just justice, though that too. What's missing is basic logic. If you want a climate mitigation strategy to work, it has to be locally appropriate and socially feasible. If it doesn't make sense to the people expected to implement it, or if it directly undermines their governance systems, then it's unlikely to deliver results. Striving for 'thick civil society' as Baker et al. (2021) encourage is a logical step.

There have been promising attempts to build out models that align with this logic. One of the earlier efforts was the Community Options Analysis and Investment Tool (COAIT) (Bonis Charancle et al., 2009), developed by Innovative Resources Management with USAID support in the 2000s. COAIT was an early attempt at hybrid planning. It blended participatory mapping and data gathering with ecological, social, and economic cost–benefit analysis. This tool helped communities understand trade-offs across different land-use options and supported them in making decisions based on their priorities. As documented in *Redeeming REDD* (Brown, 2013), it was an approach rooted in realism, not just recognition.

Whether through new tools or old principles, the bottom line remains clear: Nominal empowerment without tangible planning and management capability is a false promise. If IPLCs are to serve as anchors for sustainable climate action that NbS and Natural Climate Solutions (NCS) envision (Griscom et al., 2017), they need real leverage. That means secure tenure, meaningful participation, and the institutional infrastructure to negotiate terms that work for them and the broader climate agenda. This breakdown in implementation is not coincidental. It stems from deeper structural tensions that remain unresolved.

Despite these advances, most so-called inclusive or participatory approaches remain fundamentally consultative, with real decision-making power and control over resources still held by external agencies or donors.

Recent research on "inclusive conservation" demonstrates the value of interventions such as education, capacity building, and partnership development in protected area management and highlights the importance of empowering local communities, women, and youth (see Cebrián-Piqueras et al., 2023). Initiatives like the Inclusive Conservation Initiative (ICI), led by Conservation International and IUCN represent a partial shift away from wholly top-down models by channeling more resources and recognition to IPLCs and piloting new governance structures (see GEF, 2019; IUCN and Conservation International, 2021).

However, the ICI's primary focus remains biodiversity conservation and natural resource management, with climate mitigation integrated as a co-benefit rather than as a central objective. Its design and reporting emphasize SDG 15 and the protection of biodiversity hotspots, with climate mitigation outcomes, such as reduced CO_2 emissions, framed as additional global environmental benefits. While laudable, one might argue that the approach perpetuates the fundamental problem with the dominant conservation paradigm that applies to CMLSGS as well.

For example, as both the literature and ICI's own internal evaluation show, inclusive conservation stops at participation or delegated responsibility, without transferring genuine authority over planning and policy to IPLCs and other local actors. At the end of the day, ICI exemplifies a lingering issue driven by thought leadership and power vested in the Global North: top-down design and decision-making justified by a "consultative model" that, while respectful of Indigenous Knowledge (IK) and the crucial role IPLCs must play in conservation and climate mitigation, does not actually involve them in negotiating the terms that matter most. These are the terms that will make or break long-term sustainability, not simply project funding flows.

The CI and IUCN GEF Project Agencies will provide project assurance, including supporting project implementation by maintaining oversight of all technical and financial management aspects. They will also monitor the project's implementation and achievement of the project outputs, ensure the proper use of GEF funds, and review and approve any changes in budgets or workplans" (CI, IUCN, 2021, p. 28). In other words, decision-making and control over funding and project design of this

$68.5 million project remain with CI, IUCN, and the GEF. ICI updated its progress in delivering "$14.5 million in direct access agreements with Indigenous-led initiatives across multiple countries" (ICI, 2024). Generalized across the Global South, IP-LCs are rarely will not become positioned as equal negotiating partners in the overall architecture of landscape-level initiatives of global importance with this limited level of programming and funding. Why in 2025 is that still the case?

It is fair to ask: How long can such a top-down model be sustained, in a world where IPLCs are increasingly aware not only of their rights but also of the value of their conservation contributions to the international community?

The NGP approach proposed here as a core governance component in a much-needed paradigm shift is designed to move beyond "inclusion" toward tangible power-sharing. NGPs are not simply about inviting IPLCs into existing frameworks; they create formal, negotiated spaces where IPLCs are co-architects of strategies, with recognized rights and the capacity to shape land use and climate outcomes at every stage. In this way, NGPs address the persistent gap between "inclusion" and more genuine empowerment, offering a structural pathway to co-governance and sustainable results for people and planet.

As decades of conservation and climate programming have shown, even the most technically sophisticated or well-funded strategies falter if they are not grounded in the realities of the landscapes and communities where they are to be implemented. No climate mitigation strategy will succeed unless it fits the context, is socially feasible for those expected to carry it out, and is technically feasible within the political, economic, and institutional constraints on the ground. These three conditions – appropriateness, social feasibility, and technical feasibility – are interdependent. If anyone is ignored, the entire effort is likely to fail, either outright or through erosion of trust, legitimacy, and outcomes over time.

A Scenario for a Paradigm Shift That Will Fully Please No One: Why It May Be Timely and Work

In an ideal world that does not exist, IPLCs would receive direct, predictable, and results-based funding at scale to steward their lands for climate, biodiversity, and SDG outcomes. Such a model would build on the foundation that IPLCs manage or claim customary tenure over at least a third of the world's land, including many of its most carbon-rich and biodiverse ecosystems (Right and Resources Initiative, 2023; WRI, 2024), and would establish a plausible basis for "triple win" outcomes (see Priestley Center for Climate Future, n.d.). Where IPLC rights are secure, deforestation rates are dramatically lower, and ecological resilience is higher than in state or private hands. Yet the current reality is starkly different: Less than 2% of climate finance flows directly to IPLCs, while intermediaries and corporate actors capture most of the value, often undermining both equity and effectiveness (Rights and Resources Initiative, 2023), potentially negatively impacting on social mobility as well (Iversen et al., 2021).

The model proposed here, a multi-stakeholder, negotiated governance framework, bridges this gap. It is not about creating new bureaucracies or rent-seeking structures but about enabling IPLCs to act as co-architects of land-use strategies. This should be supported by direct funding streams from governments, private sector actors (especially those with Beyond Value Chain Mitigation commitments), and philanthropies. This approach is not merely aspirational; it is a pragmatic necessity. IPLCs influence resource use across at least one-third of the globe, including critical carbon sinks like the Amazon and Congo Basin. Without their leadership, the world has no credible pathway to limiting warming to 1.5°C or halting biodiversity loss (World Bank, 2024).

Where agreed through NGPs, intermediaries capturing the bulk of climate funding should be replaced with IPLC-managed financing mechanisms. Models such as the Shandia Platform of the Global Alliance of Territorial Communities, which supports 35 million IPLC global members, and the Mesoamerican Territorial Fund, or direct transfer models like Cool Earth, which delivers flexible, unconditional support directly to forest communities, offer grounded, innovative, representative funding approaches. These approaches aim to ensure that at least 50% of climate finance reaches IPLCs, reducing the dilution of funds through costly intermediaries. They represent an important springboard for expanding IPLC funding in the future, especially for cases where direct funding to Indigenous Peoples through the Shandia Platform is feasible for donors and governments.

The GEF has found that such direct access models can reduce transaction and overhead costs by up to 30% when compared to traditional donor-NGO systems. This is a factor that must be weighed strategically in the current context of deep Overseas Development Assistance (ODA) budgetary cutbacks in 2025.

Kenya's 2024 Climate Change (Carbon Markets) Regulations provide an emerging example of a major government establishing a codified framework for carbon offset projects "undertaken on community lands." The regulations mandate that:

- Land-based carbon projects allocate at least 40% of their aggregate annual profits to community development; and
- Non-land-based carbon projects allocate at least 25% of profits to the same.

While this falls short of full revenue sharing, it represents a legally enforceable step toward more equitable IPLC participation in climate mitigation benefits and demonstrates the kind of national framework that NGPs, described in more detail below, can learn from, build upon.

To maintain accountability and ensure learning, payments should be linked to verifiable outcomes co-defined by IPLCs. Metrics such as forest cover maintenance, land-use practices, species recovery, land conflicts, and sustainable agroforestry yields must be favored over MRV tied solely to carbon metrics, which prioritize values keen to the Global North rather than those of IPLCs and governments in the Global South. However, where IPLCs choose on their own volition to enter carbon markets and choose to negotiate their terms for engagement, they clearly should retain that right. That is but one facet of empowerment.

MRV-driven legitimacy must give way to social license anchored in negotiated governance. That means replacing symbolic participation with genuine co-governance, moving from indirect benefits from future carbon sales to direct negotiation of land-use terms, and recognizing that without IPLC leadership, sustainability will remain elusive.

This negotiated, multi-stakeholder approach is unlikely to fully please any single actor. Yet, in its realism and balance, it offers the most credible path forward – a paradigm shift that is both timely and necessary for achieving durable, equitable climate outcomes. It is time the international community truly walked the talk on co-production and collaboration.

Negotiated Governance Platforms

NGPs are envisioned as structured, tripartite platforms at the jurisdictional level where IPLCs, governments, and private actors negotiate governance arrangements over sustainable land-use plans, benefit-sharing, Scope 3 relevant supply chain management, and safeguards. The Suruí Forest Carbon Project in Brazil, while often cited as a pioneering example of IPLC-led negotiation, also illustrates the challenges and limitations of this approach.

The Paiter-Suruí people negotiated a Memorandum of Understanding guaranteeing community control over carbon revenues and project design through iterative dialogue with technical partners such as U.S. NGO Forest Trends, market intermediaries, and, to a lesser extent, government actors. All acting chiefs signed the cooperation agreement, and the proposal was debated and approved by the Forum of the Clans, demonstrating a negotiated, community-driven approach to project terms and benefit sharing. Crucially, the process was participatory and iterative, with FPIC achieved through multiple rounds of community meetings and scenario analysis (Surui Forest Carbon Project, 2014).

However, it is equally important to acknowledge that the Suruí project has faced legitimate criticisms. Despite its pioneering status, the project encountered difficulties related to market volatility, technical complexity, and the burden of ongoing monitoring and reporting. Some community members expressed concerns about the distribution of benefits, the sustainability of revenues, and the challenges of maintaining long-term engagement with external partners. These issues highlight the risks of relying on carbon projects as a primary pathway and reinforce the book's core argument: Carbon markets should be an option for IPLCs, not an obligation or default solution. The Suruí experience underscores the need for robust capacity building, transparent negotiation, and ongoing adaptive management – elements that are central to the negotiated governance paradigm advocated here (see also Rights and Resources Initiative, 2023; Sills et al., 2014).

Evidence from the field continues to mount on the perceived role IPLCs must play in climate and conservation. These models show that when IPLCs are empowered as co-governors, outcomes improve for people and planet alike. For example, the Acorn Initiative (Rabobank) in Latin America allows smallholder farmers to retain 75% of carbon credit revenues, using blockchain technology for verification and scaling to 150,000 farmers (World Bank, 2024). Kenya's Community Land Act reduced land disputes by 40% and increased IPLC participation in climate initiatives (Kenya Gazette, 2024).

Yet the Suruí case and others remind us: Negotiated governance is not a panacea. It is a process fraught with complexity, trade-offs, and the risk of disappointment. Market volatility, technical barriers, and the ever-present danger of elite capture or benefit-sharing disputes are real (Worsdell & Sambhav (2020) . But these are not arguments against negotiation. There are arguments for doing it better, with greater transparency, stronger safeguards, and a relentless focus on IPLC agency and choice.

Graphic 7a is a visual synthesis of institutional architecture rooted in negotiated governance, tenure security, and capacity building, enabling IPLC co-governance and appropriate STAM for credible climate outcomes. The framework illustrates what is required for socially feasible mitigation under the CMLSGS paradigm.

The graphic synthesizes the chapter's core argument: Negotiated governance, tenure security, and capacity building must be systematized into national strategies to produce sustainable outcomes. This graphic is not a blueprint but a guidepost for restructuring the institutional relationships needed to overcome the structural limitations discussed throughout this chapter.

An Illustrative Roadmap for NGPs in CMLSGS in Diverse Settings

Establishing NGPs is a multi-step process that begins with securing a broad social license at the national level and proceeds through structured negotiation, legal codification, implementation, and review. Table 7.1 summarizes the core steps, key stakeholder roles, and critical safeguards needed to ensure legitimacy and effectiveness at each stage. While not meant to be an inflexible template, the following roadmap can help those open, willing, and able to innovate the key planning and governance elements of the paradigm to achieve sustainable CMLSGS.

Obtaining a genuine social license may arguably be the most important step in the entire NGP process, for without broad-based legitimacy and trust, none of the subsequent actions in the roadmap are likely to be achievable in practice. If stakeholders do not perceive the platform as legitimate, inclusive, and responsive to their rights and interests, negotiations risk stalling, agreements may lack buy-in, and implementation will almost certainly face resistance or failure (FAO, 2006; Rainforest Foundation UK, 2024). Securing this foundation is therefore not just a procedural formality but the prerequisite for every other step in effective, negotiated governance.

The current paradigm for CMLSGS is no longer logical, socially acceptable, politically stable, or financially defensible. Its reliance on technocratic control and market abstraction has failed to deliver convincing outcomes. What this book has argued instead is that negotiated governance – pluralistic, adaptive, and grounded in the lived realities of land stewards – offers not just a better path but also the only one still open if measurable-versus rhetorical – CMLSGS is the goal. Credible, lasting climate solutions will depend not on external blueprints, but on what happens when those who customarily own the land are empowered, at a minimum, to co-govern its future.

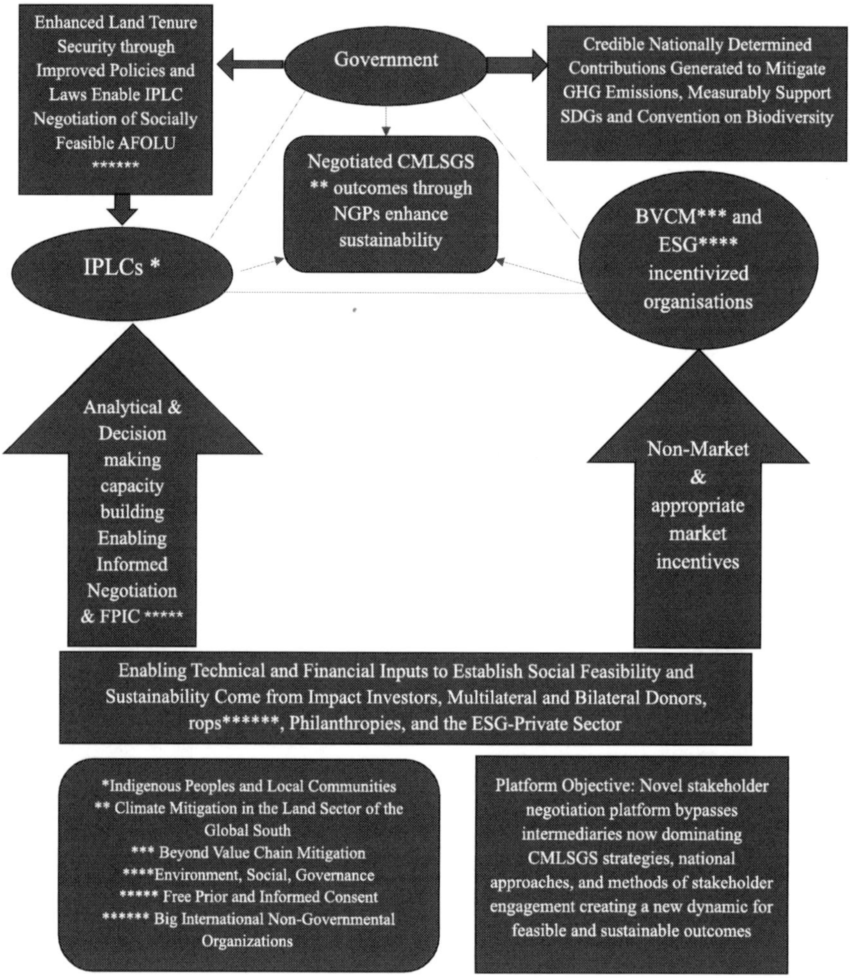

Figure 7.1 Framework for Socially Feasible Climate Mitigation in the Global South

Mutirão for the Future: Negotiated Collaboration and IPLC Leadership

Brazil's call for a global mutirão – a collective, bottom-up mobilization for climate action, now championed at COP30 in Belém – captures a spirit of collaboration that this book both echoes and proposes a pathway to operationalize (see Second Letter from the COP30 Presidency, May 2025; COP30 Newsroom, 2025). Yet, as the evidence and experience across the Global South show, consensus on "bottom-up" action is not enough. Durable, equitable outcomes require structured, negotiated collaboration. IPLC leadership and community representation must anchor all structured negotiation processes and be supported by governments, the private

Table 7.1 NGP Operationalization Roadmap

Step	*Key Actions and Decisions*	*Stakeholders*	*Safeguards & Notes*
0. Social License	Map stakeholders; catalyze national dialogue; agree on values, tenure, incentives, and terms; and document "social contract."	IPLCs, government, private sector, civil society, donors	IPLC rights, tenure, incentives, clear terms, and early capacity needs assessment
1. Define Scope and Goals	Set mandate, objectives, land tenure/incentive priorities, FPIC requirements.	All	FPIC, gender equity, tenure security, incentives; capacity plan
2. Stakeholder Engagement	Identify/map all parties; pre-negotiation consultations; **initiate FPIC/negotiation capacity building for IPLCs**.	All	Inclusive mapping, tenure recognition, tailored incentives, FPIC support
3. Governance Structure	Form a committee; set rules (incl. FPIC compliance); working groups; codify engagement terms.	IPLCs, government, private, civil society	IPLC/gender roles, tenure expertise, enforceable terms, ongoing FPIC training
4. Negotiation and Agreement	Joint fact-finding; negotiate tenure, incentives, FPIC; draft agreement.	All	Legal review, public disclosure, grievance, IPLC technical/legal support
5. Legalization and Resources	Formalize agreement (incl. tenure/FPIC); allocate resources; binding terms.	Government, private, IPLCs	Compliance-based disbursement, tenure security, FPIC resources
6. Implemen-tation and Adaptation	Assign roles; monitor (incl. FPIC); adapt as needed.	All	Participatory monitoring, feedback, and continuous IPLC capacity building
7. Review and Accountability	Independent audits (incl. FPIC/capacity); public reporting; periodic review.	All external evaluators	Transparency, IPLC/civil society triggers, review FPIC effectiveness

References: TNC Human Rights Guide (2024); Accountability Framework (2020); UNPFII/IFAD (2005); IUCN and Conservation International (2022); NatureMetrics (2025).

sector, and civil society (see Rights and Resources Initiative, 2023; World Bank, 2021). Tools like the Nature Tech Collective and Impact Lab's (2024) Framework or Farmonaut's Satellite Farm Advisor (2024) could, in theory, align structured technical support to IPLC-led processes to enhance prospects for success.

This book has argued that top-down models and externally imposed frameworks have failed to deliver sustainable or just outcomes. Where IPLCs have secure tenure and a real seat at the table, results for climate, biodiversity, and livelihoods are better (see Garnett et al., 2018; World Resources Institute, 2024). Achieving this outcome demands more than consultation or tokenism. It requires systematic investment in tenure security, capacity building, and the creation of NGPs – locally adapted, multi-level spaces for co-designing climate and land strategies (see FAO, 2006; UNDP FPIC Guidance, 2022). These fit under a "contributions approach" to finance valuable projects (Romm et al., 2025) based on technical and social feasibility.

The mutirão ethos is a powerful rallying cry, but only through negotiated, collaborative programming – anchored in IPLC leadership and government partnership – can we move beyond persistent and implausible rhetoric to far more credible and lasting outcomes for people and planet.

References

Accountability Framework Initiative (AFI). (2020). The Accountability Framework: Core Principles and Operational Guidance. *AFI,* Washington DC. https://accountability-framework.org/the-accountability-framework/core-principles/

Agrawal, A., & Chhatre, A. (2006). Explaining success on the commons: Community forest governance in the Indian Himalaya. *World Development, 34*(1), 149–166. https://www.sciencedirect.com/science/article/abs/pii/S0305750X05001889

Angelsen, A., Martius, C., De Sy, V., Duchelle, A. E., Larson, A. M., & Pham, T. T. (Eds.). (2018). *Transforming REDD+: Lessons and New Directions.* CIFOR.

Artelle, K. A., Zurba, M., Bhattacharyya, J., Chan, D. E., Brown, K., Housty, J., & Moola, F. (2019). Supporting resurgent Indigenous-led governance: A nascent mechanism for just and effective conservation. *Biological Conservation*, 240, 108284. https://www.sciencedirect.com/science/article/pii/S0006320719307803

Baker, S., Ayala-Orozco, B., & García-Frapolli, E. (2021). The role of civil society organisations in climate change governance: Lessons from Quintana Roo, Mexico. *Journal of the British Academy*, 9(s10), 99–126. https://www.thebritishacademy.ac.uk/publishing/journal-british-academy/9s10/civil-society-organisations-climate-change-governance/

Bebbington, A., & Humphreys Bebbington, D. (2011). An Andean avatar: Post-neoliberal and neoliberal strategies for securing the unobtainable. *New Political Economy*, 16(1), 131–145. https://commons.clarku.edu/faculty_geography/486/

Berkes F, Colding J, Folke C. (2000). Rediscovery of traditional ecological knowledge as adaptive management. *Ecological Applications*. 10, 1251–1262. https://www.researchgate.net/publication/235624698_Rediscovery_of_Traditional_Ecological_Knowledge_as_Adaptive_Management

Bonis-Charancle, J. M., Tiani, A. M., Brown, M., Akwah Neba, G., Mogba, Z., Lescuyer, G., Warne, R., & Greenberg, B. (2009). Strengthening local analytical capability: Community options analysis and investment. In M. C. Diaw, T. Aseh, & R. Prabhu (Eds.), *In Search of Common Ground: Adaptive Collaborative Management in Cameroon* (pp. 275–300). CIFOR. https://www.researchgate.net/publication/302374627_Linking_the_Community_Options_Analysis_and_Investment_Toolkit_COAIT_ConsensysR_and_Payment_for_Environmental_Services_PES_A_Model_to_Promote_Sustainability_in_African_Gorilla_Conservation

Boucher, D., et al. (2014). *Deforestation Success Stories: Tropical Nations Where Forest Protection and Reforestation Policies Have Worked.* Union of Concerned Scientists. https://www.ucs.org/resources/deforestation-success-stories

Brown, M. I. (2013). *Redeeming REDD: Policies, Incentives and Social Feasibility for Avoided Deforestation*. Routledge.
Carbon Market Watch. (2023). *Error log: Exposing the methodological failures of REDD+ forestry projects*. https://carbonmarketwatch.org/publications/error-log-exposing-the-methodological-failures-of-redd-forestry-projects/
CGIAR Research Program on Climate Change, Agriculture and Food Security (CCAFS). (2017). *"Triple wins" or "triple faults"?: Unpacking the equity implications of climate-smart agriculture*. https://ccafs.cgiar.org/index.php/news/triple-wins-or-triple-faults-unpacking-equity-implications-climate-smart-agriculture
Cebrián-Piqueras, M. A., Palomo, I., Lo, V. B., López-Rodríguez, M. D., Filyushkina, A., Fischborn, M., Raymond, C. M., & Plieninger, T. (2023). Leverage points and levers of inclusive conservation in protected areas. *Ecology and Society,* 28(4), Article 7. https://research.vu.nl/en/publications/leverage-points-and-levers-of-inclusive-conservation-in-protected
ClientEarth. (2023). Community-Based Forest Management: Legal Frameworks in Five Congo Basin Countries. Retrieved from https://www.clientearth.org/latest/documents/community-based-forest-management-legal-frameworks-in-five-congo-basin-countries/
Climate Action Network. (2024). ECO 4, COP29: It isn't just about getting funds – it's a matter of justice. https://climatenetwork.org/resource/eco-4-cop29/
Climate Investment Funds (CIF). (2023). Annual Report 2023: *Bolder Ambitions, More Urgency, Unflinching Determination*. CIF, Washington, DC. https://www.cif.org/sites/cif_enc/files/knowledge-documents/cif_annual_report_2023.pdf
Conservation International (CI) & International Union for Conservation of Nature (IUCN). (2021). Inclusive Conservation Initiative: Enhance Indigenous Peoples' and Local Communities' (IPLCs) efforts to steward land, waters and natural resources to deliver global environmental benefits. Retrieved from https://www.thegef.org/projects-operations/projects/10404
COP30 Newsroom. (2025). Brasil launches global call for a "mutirão" against climate change. https://cop30.br/en/news-about-cop30-amazonia/brazil-launches-global-call-for-a-mutirao-against-climate-change
Cultural Survival. (2022). Despite historic participation, Indigenous Peoples are again sidelined in major decisions at COP27. *Cultural Survival News*. https://www.culturalsurvival.org/news/despite-historic-participation-indigenous-peoples-are-again-sidelined-major-decisions-cop27
Dawson, N. M., Coolsaet, B., Sterling, E. J., Loveridge, R., Gross-Camp, N. D., Wongbusarakum, S., Sangha, K. K., Scherl, L. M., Phuong Phan, H., Zafra-Calvo, N., Lavey, W. G., Byakagaba, P., Idrobo, C. J., Chenet, A., Bennett, N. J., Mansourian, S., & Rosado-May, F. J. (2024). Is it just conservation? A typology of Indigenous Peoples' and local communities' roles in conserving biodiversity. *One Earth*, 7(6), 1007–1021.
Drawdown. (2023). Triple win: Climate change, poverty, and biodiversity. https://drawdown.org/insights/triple-win-climate-change-poverty-and-biodiversity
Daytec-Yañgot, C. L. (2021, September). *FPIC: A Shield or Threat to Indigenous Peoples' Rights? In Indigenous Peoples Experiences on Free, Prior and Informed Consent: A Collection of Case Studies (Philippines)*. UN-REDD Programme. https://www.un-redd.org/sites/default/files/2021-09/Indigenous%20Peoples%20Experiences%20on%20Free%2C%20Prior%20and%20Informed%20Consent%3A%20A%20Collection%20of%20Case%20Studies%28Philippines%29.pdf
Dudine, K., & Szoke-Burke, S. (2020). Government Briefing: Incorporating Free, Prior, and Informed Consent into Investment Approval Processes. *Columbia Center on Sustainable Investment,* New York, USA. https://ccsi.columbia.edu/sites/ccsi.columbia.edu/files/content/docs/Briefing-FPIC-and-investment-approval-July-2020%20(1).pdf
FAO. (2023). *The State of Food and Agriculture 2023 – Revealing the true cost of food to transform agrifood systems*. Rome. https://doi.org/10.4060/cc7724en
Farmonaut. (2024). The satellite farm advisor. https://farmonaut.com

Food and Agriculture Organization of the United Nations (FAO). (2006). Land tenure alternative conflict management (FAO Land Tenure Manuals No. 2). Rome: FAO. https://globalprotectioncluster.org/sites/default/files/2023-04/Land_Tenure_Alternative_Conflict_Management_2006_EN.pdf

Food and Agriculture Organization of the United Nations (FAO) & United Nations Environment Programme (UNEP). (2020). The state of the world's forests 2020: Forests, biodiversity and people. Rome: FAO.

Forest Trends. (2014). *The Surui Forest Carbon Project: Building the Foundation for REDD+ in Indigenous Territories*. Washington, D.C.: Forest Trends.

Friedman, M. (1982). *Capitalism and Freedom* (Revised ed.). University of Chicago Press.

Fursman, L. (2025). *The Climate Paradox: Why We Need to Reset Action on Climate Change*. Tony Blair Institute for Global Change. https://institute.global/insights/climate-and-energy/the-climate-paradox-why-we-need-to-reset-action-on-climate-change

Future Agricultures. (2013). Climate compatible development: The politics of "triple wins" and trade-offs. https://www.future-agricultures.org/blog/climate-compatible-development-the-politics-of-triple-wins-and-trade-offs/

Galvin, S. (2025, April 10). Elon Musk touts $150B in savings in FY2026 – after claiming DOGE could cut $1T. *New York Post*. https://nypost.com/2025/04/10/us-news/elon-musk-touts-150b-in-savings-in-fy2026-after-claiming-doge-could-cut-1t/

Garnett, S. T., Burgess, N. D., Fa, J. E., Fernández-Llamazares, Á., Molnár, Z., Robinson, C. J., . . . & Leiper, I. (2018). A spatial overview of the global importance of Indigenous lands for conservation. *Nature Sustainability*, 1(7), 369–374.

Global Environment Facility (GEF) (2019). Inclusive conservation initiative – project identification form. https://www.thegef.org/sites/default/files/web-documents/10404_BD_PIF.pdf

Global Witness. (2024). The Cerrado crisis: Brazil's deforestation frontline. Published December 29, 2024. https://www.globalwitness.org/en/campaigns/forests/the-cerrado-crisis-brazils-deforestation-frontline/

Global Witness. (2025). 5 things to know about the Tropical Forest Forever Facility. Published May 28, 2025. https://globalwitness.org/en/campaigns/forests/5-things-to-know-about-the-tropical-forest-forever-facility/

Griscom, B. W., et al. (2017). Natural climate solutions. *Proceedings of the National Academy of Sciences*, 114(44), 11645–11650.

Henneberger, B. (2025). The Indigenous Peoples and local communities' road to COP30 at Belém. *Environmental Defense Fund Blog*. https://blogs.edf.org/climate411/2025/01/29/indigenous-peoples-local-communities-road-to-cop30/

Inclusive Conservation Initiative. (2021, May). Environmental and social management framework for Component 1: "Local IPLC action to deliver global environmental benefits (GEB): Established on-the-ground projects led by IPLC organizations" (Prepared by Asesoramiento Ambiental Estratégico, AAE). https://iucn.org/sites/default/files/2022-09/esmf-component-1-inclusive-conservation-gef.pdf

Inclusive Conservation Initiative. (2024). Phase Two Report: A focus on inclusive finance. Washington, DC, USA: Inclusive Conservation Initiative. https://www.thegef.org/sites/default/files/documents/2024-08/ICI-Phase-Two-Report-2024.pdf

Intergovernmental Science-Policy Platform on Biodiversity and Ecosystem Services. (2019). Global assessment report on biodiversity and ecosystem services. Brondízio, E. S., Settele, J., Díaz, S., & Ngo, H. T. (Eds.). IPBES Secretariat, Bonn, Germany. https://doi.org/10.5281/zenodo.3831673

International Initiative for Impact Evaluation (3ie). (2024). Understanding the evidence on climate change, biodiversity and livelihoods. https://www.3ieimpact.org/research/understanding-evidence-climate-change-biodiversity-and-livelihoods

Intergovernmental Panel on Climate Change. (2022). *Climate change 2022: Mitigation of climate change*. Contribution of Working Group III to the Sixth Assessment Report of the Intergovernmental Panel on Climate Change. https://www.ipcc.ch/report/ar6/wg3/

Iversen, V., Krishna, A., & Sen, K. (2021). What do we need to know about social mobility in the Global South? (WIDER Research Brief 2021/5). Helsinki: UNU-WIDER.
Jalali, M. S. (2025). WHO and USAID budget cuts hurt the US. *Science*, 388(6747), 500–501. https://doi.org/10.1126/science.adw8264
Kenya Gazette. (2024). *The Climate Change (Carbon Markets) Regulations, 2024*. Nairobi: Government of Kenya.
Lofts, K., Sarmiento Barletti, J. P., & Larson, A. M. (2021). Lessons towards rights-responsive REDD+ safeguards from a literature review (Working Paper 280). Bogor, Indonesia: CIFOR.
Malig, M. L., & Solón, P. (2025). TFFF: A False Solution for Tropical Forests. *Global Forest Coalition & Fundación Solón*. Retrieved from https://globalforestcoalition.org/report-tfff/
NatureMetrics. (2025). Report interpretation guide. *NatureMetrics,* London. https://www.naturemetrics.com/report-interpretation-guide
Nature Tech Collective & ImpactLabs. (2024). Developing the Nature Tech Taxonomy Framework. https://www.naturetechcollective.org/stories/nature-tech-taxonomy-framework
Nepstad, D., et al. (2019). Pathways for recent Cerrado soybean expansion: Extending the soy moratorium and implementing integrated crop livestock systems with soybeans. *Environmental Research Letters*, 14(5), 055001. https://iopscience.iop.org/article/10.1088/1748-9326/aafb85
Pörtner, H. O., Scholes, R. J., Agard, J., Archer, E., Arneth, A., Bai, X., Barnes, D., Burrows, M., Chan, L., Cheung, W. L., Diamond, S., Donatti, C., Duarte, C., Eisenhauer, N., Foden, W., Gasalla, M., Handa, C., Hickler, T., Hoegh-Guldberg, O., ... Ngo, H. T. (2021). IPBES-IPCC co-sponsored workshop report on biodiversity and climate change. Intergovernmental Science-Policy Platform on Biodiversity and Ecosystem Services (IPBES) and Intergovernmental Panel on Climate Change (IPCC).
Priestley Centre for Climate Futures. (n.d.). Triple-win adaptations: Tackle climate change, land degradation and biodiversity loss. University of Leeds. https://climate.leeds.ac.uk/research/understanding-risk-to-develop-a-resilient-world/triple-win-adaptations-tackle-climate-change-land-degradation-and-biodiversity-loss/
Rainforest Foundation UK. (2024, October 30). Congolese organisations welcome cancellation of oil block auction, warn against new tender. https://www.rainforestfoundationuk.org/congolese-organisations-welcome-cancellation-of-oil-block-auction-warn-against-new-tender/
Rights and Resources Initiative. (2020). *Rights-Based Conservation: The Path to Preserving Earth's Biological and Cultural Diversity?* Washington, DC: Rights and Resources Initiative.
Rights and Resources Initiative. (2023). Who owns the world's land? Global state of Indigenous, Afro-descendant, and local community land rights recognition from 2015–2020 (2nd ed.). *Rights and Resources Initiative*. https://2023report.rightsandresources.org/global-achievements/who-owns-the-worlds-land#:~:text=The%20report%20analyzes%20the%20legal%20frameworks%20of%2073,and%20local%20community%20ownership%20between%202015%20and%202020
Rogelj, J., et al. (2021). Net-zero emission targets for major emitting countries consistent with the Paris agreement. *Nature Communications*, 12(1), 2047.
Romm, J., Lezak, S., & Alshamsi, A. (2025). Are Carbon Offsets Fixable? *Annual Review of Environment and Resources*, 50, 649–680. https://www.annualreviews.org/content/journals/10.1146/annurev-environ-112823-064813
Rose, S., Collinson, E., & Kalow, J. (2017). Working itself out of a job: USAID and smart strategic transitions. *Center for Global Development*. Retrieved from: https://www.cgdev.org/sites/default/files/working-itself-out-job-usaid-and-smart-strategic-transitions.pdf
Sarmiento Barletti, J. P., & Larson, A. M. (2017). *Rights abuse allegations in the context of REDD+ readiness and implementation: A preliminary review and proposal for moving forward* (p. 190). CIFOR InfoBrief.
Schleicher, J., Peres, C. A., Amano, T., Llactayo, W., & Leader-Williams, N. (2017). Conservation performance of different conservation governance regimes in the Peruvian Amazon. *Scientific Reports*, 7(1), 11318. https://doi.org/10.1038/s41598-017-10736-w

Sills, E. O., Atmadja, S. S., de Sassi, C., Duchelle, A. E., Kweka, D. L., Resosudarmo, I. A. P., & Sunderlin, W. D. (Eds.). (2014). *REDD+ on the ground: A case book of subnational initiatives across the globe.* CIFOR.

SOMO & Kenya Human Rights Commission. (2023). Offsetting human rights: Sexual abuse and harassment at the Kasigau Corridor REDD+ Project in Kenya. *SOMO & KHRC.* https://www.somo.nl/offsetting-human-rights/

Streck, C. (2024, April 22). Quote from: What's the TFFF? A forest finance tool "like no other" shows potential. *Mike Digirolamo Interview.* https://news.mongabay.com/podcast/2024/12/whats-the-tfff-a-forest-finance-tool-like-no-other-shows-promise-pitfalls/

Sunderlin, W. D., Larson, A. M., Duchelle, A. E., Resosudarmo, I. A. P., Huynh, T. B., Awono, A., & Dokken, T. (2014). How are REDD+ proponents addressing tenure problems? Evidence from Brazil, Cameroon, Tanzania, Indonesia, and Vietnam. *World Development, 55,* 37–52. https://doi.org/10.1016/j.worlddev.2013.01.013

Suckall, N., Stringer, L. C., & Tompkins, E. L. (2015). Presenting triple-wins? Assessing projects that deliver adaptation, mitigation and development co-benefits in rural Sub-Saharan Africa. *Ambio*, 44, 34–41.

The Nature Conservancy (TNC). (2023). *Human Rights Guide for Working with Indigenous Peoples and Local Communities.* Available at: https://www.tnchumanrightsguide.org

The Nature Conservancy. (2024). TNC Human Rights Guide. Available at: https://www.tnchumanrightsguide.org

United Nations Development Programme. (2014). *Governance for Sustainable Development: Integrating Governance in the Post-2015 Development Framework.*

United Nations Development Programme (UNDP). (2022). Supplemental Guidance: Frequently Asked Questions on Applying Free, Prior, and Informed Consent (FPIC) for NCE-supported Projects. *UNDP Social and Environmental Standards (SES).* https://ses-toolkit.info.undp.org/standard-6

United Nations Environment Programme. (2022). *Emissions Gap Report 2022: The Closing Window – Climate Crisis Calls for Rapid Transformation of Societies.* United Nations Environment Programme.

United Nations Framework Convention on Climate Change. (2024). Local Communities and Indigenous Peoples Platform (LCIPP). *United Nations Framework Convention on Climate Change.* https://lcipp.unfccc.int/?utm_source=chatgpt.com

United Nations. (2023). The Sustainable Development Goals Report 2023: Special Edition. Department of Economic and Social Affairs. https://unstats.un.org/sdgs/report/2023/

United Nations Permanent Forum on Indigenous Issues & International Fund for Agricultural Development. (2005).Technical workshop report: Indigenous Peoples, development with identity, indigenous institutions and governance structures. *UNPFII/IFAD.* https://www.un.org/esa/socdev/unpfii/documents/fifthsession_tk_en.pdf

World Bank. (2021). *Report on Land Tenure Security.* Washington, D.C.: World Bank.

World Bank. (2024). *Climate-Smart Agriculture: From Knowledge to Implementation.* Washington, DC: World Bank Group. https://www.worldbank.org/en/results/2024/12/05/climate-smart-agriculture-from-knowledge-to-implementation

World Economic Forum. (2024). Private climate finance: 4 things to consider. https://www.weforum.org/stories/2024/04/private-climate-finance-4-things-you-need-to-know/

World Rainforest Movement. (2025, April 22). "Worst conservation fund ever." Published April 22, 2025. https://www.wrm.org.uy/bulletin-articles/worst-conservation-fund-ever

World Resources Institute. (2024, November). *Protecting Biodiversity Hinges on Securing Indigenous and Community Land Rights.* https://www.wri.org/insights/indigenous-and-local-community-land-rights-protect-biodiversity

Index

Note: Numbers in **bold** indicate a table. Numbers in *italics* indicate a figure. Numbers that are underlined indicate a box.

Made in the USA
Monee, IL
12 November 2026

0c156666-b280-42fc-9014-db41ee984d70R01